Series Title
Engineering Dynamics and Vibration

Edited By

Cho W. S. To

*Department of Mechanical and
Materials Engineering, University of Nebraska
Lincoln, Nebraska
U.S.A.*

Volume Title
Vibration and Nonlinear Dynamics of Plates and Shells
Applications of Flat Triangular Finite Elements

Authored By

Meilan Liu

Department of Mechanical Engineering
Lakehead University, Thunder Bay
Ontario
Canada

&

Cho W. S. To

Department of Mechanical and
Materials Engineering, University of Nebraska
Lincoln, Nebraska
U.S.A.

CONTENTS

FOREWORD

Engineering Dynamics and Vibration are two foundation areas in many engineering fields. They are fundamental to the analysis and design of many dynamic engineering systems. The present volume, *Vibration and Nonlinear Dynamics of Plates and Shells: Applications of Flat Triangular Finite Elements* is a timely and unique addition to the literature in the two foundation areas in engineering. The authors, Professors Meilan Liu and C.W. Solomon To, have a combined experience of more than fifty years in the engineering dynamics and vibration analysis of plate and shell structures. The present volume has included the two foundation areas in a single book that applies the lower order flat triangular shell finite elements originated from their early research. Its main and important feature is the combination of vibration and nonlinear dynamics of plates and shells in a relatively comprehensive treatment employing the finite element method. Another feature of the present volume is the treatment of boxed or cell structures. It is believed that anyone working in the analysis and design of dynamic engineering systems will find it informative and an excellent reference.

Neil Popplewell

Professor of Mechanical Engineering
University of Manitoba
Canada

Series Preface

The fields of engineering dynamics and vibration have advanced and expanded at an extremely impressive rate due, perhaps, to their high demand in applications in modern technologies. The main objectives of this series are three folds. The first objective is to be complimentary to existing books and handbooks in the fields of engineering dynamics and vibration. The second objective is to provide a common and single venue for the publication of books in both engineering dynamics and vibration fields. Books in the emerging area of engineering dynamics and vibration of nano-structural systems and devices are included. The third objective of the present series is to provide books suitable for use by advanced undergraduates and post-graduate level engineering students, research engineers, and applied scientists.

The series aims at keeping abreast of the modern developments and applications in the fields. Whenever new areas of development and application arise it is the intent of this series to invite leaders in the field to publish their work.

Cho W. S. To
Department of Mechanical and
Materials Engineering, University of Nebraska
Lincoln, Nebraska
U.S.A.

Volume Preface

The germ of this eBook was grown from the interests of the authors in engineering vibration and dynamics. The theoretical background and computational techniques adopted throughout this eBook were based on part of the doctoral degree thesis of the first author. While the fields of computational engineering dynamics and vibration are vast, and their applications have virtually no limits, the scope of the present eBook is confined to vibration and nonlinear dynamics of plates and shells. For computational studies, the versatile finite element method alone provides a multitude of impressive publications, such as the pioneered work, *Finite Element Handbook* published in 1987 by McGraw-Hill (Editor-in-Chief, H. Kardestunder and Project Editor, D. H. Norrie). Subsequently, there are various handbooks in finite element methods available in the literature. Thus, the subject matter and topics included in the present eBook are focused on the vibration and nonlinear dynamics aspects of plate and shell structures. While finite element analysis of plates and shells is generally regarded as a mature technology it seems that no single book that covers both vibration and nonlinear dynamics by applying the finite element method is currently available. Consequently, the present volume is a modest attempt to provide such a book, albeit a relatively limited one. The particular shell finite elements employed in the computational studies reported in this book are the mixed formulation based lower order flat triangular shell finite elements.

The present book has nine chapters. The brief introduction is included in Chapter 1. Chapter 2 is concerned with the theoretical background for the vibration analysis of plates and shells. In particular, the mixed formulation based three-node flat triangular shell elements are presented in this chapter. Vibration analysis of plate structures is considered in Chapter 3. In the latter the square, circular, and skew plates as well as membrane are treated. Vibration analysis of shells with single curvature is presented in Chapter 4 in which cylindrical panel with rectangular and trapezoidal projections, Scordelis-Lo roof, and cylindrical shell clamped at both ends with its effect of aspect ratio are included. Chapter 5 is concerned with the vibration analysis of shells of double curvatures. These structures include the spherical caps, spherical panel of square projection, hemispherical panel, and clamped hemispherical shell. Chapter 6 deals with the vibration analysis of box structures. Single-cell and double-cell box structures are studied.

Chapter 7 provides the theoretical development for the nonlinear dynamic analysis of plate and shell structures. The focus is on the mixed formulation based three-node flat triangular shell elements for nonlinear dynamic analysis. Aside from presenting the steps in the derivations of the consistent element stiffness and mass matrices, constitutive relations of elastic materials and elasto-plastic materials with isotropic strain hardening, yield criterion, return mapping, configuration and stress updating strategies, and numerical algorithms are presented and discussed. Nonlinear dynamics of flat-surface structures are treated in Chapter 8. The cantilevered, circular, and square plates under uniform pressures, rectangular plate subjected to a center load, and a cubic tube under internal and external pressures are considered in this chapter. Chapter 9 is concerned with the nonlinear dynamics of curved-surface structures. The cylindrical panel under a central point load and under a uniform pressure, hemispheres with and without a central hole and under alternating point loads, clamped and hinged spherical caps subjected to apex point loads and under pressures are examined in this chapter. Finally, it should be mentioned that no attempt has been made to include the important subject of chaotic dynamics of plate and shell structures applying the lower order flat triangular shell finite elements.

ACKNOWLEDGEMENT

There is none to declare.

CONFLICT OF INTEREST

We, the authors, confirm that there is no conflict of interest in regard to contents of this book.

Meilan Liu
Department of Mechanical Engineering
Lakehead University
Thunder Bay
Ontario
Canada

Cho W. S. To
Department of Mechanical and
Materials Engineering
University of Nebraska
Lincoln, Nebraska
U.S.A.

2

Send Orders for Reprints to reprints@benthamscience.net
Vibration and Nonlinear Dynamics of Plates and Shells, 2014, 3-5   **3**

CHAPTER 1

Introduction

Abstract: This chapter consists of three sections. Objectives and scope of the book are given in Section 1.1. Section 1.2 outlines the organization. Notes on computer programming are included in the last section.

Keywords: Vibration analysis, nonlinear dynamics, plates, shells, box structures, mixed formulation, finite element method, flat triangular shell elements.

1.1. OBJECTIVES AND SCOPE

There are numerous examples of engineering structures that are composed of shell segments, such as the body of an airplane, the hull of a ship or submarine, the roof of a domed structure, a pressure vessel, and so on. A shell structure, in general, refers to a body with one dimension much smaller than the other two. That is, its thickness is much smaller than the size of the curved mid-surface. The mid-surface can be of single curvature and double curvature. Examples of former are cylinders and cones. The latter may include spherical caps, for example. Special cases of shell structures include, plates whose mid-surfaces are flat, and beams whose length is much larger than the width and thickness.

The curvatures provide shells with significant advantages over plates and beams, making shell structures perhaps the most efficient light weight structures in terms of load-carrying capacity. However, the curvatures also pose challenges for the modeling and analysis of shell structures. As Ref. [1.1] pointed out, since the mid-1960s, "the published literature on modeling of plates and shells in the linear and non-linear regimes and their application to dynamic or vibration analysis of structures has grown extensively. There has been a tremendous interest on the part of researchers with sufficiently large amount of resources devoted to the subject, and there continues to be innovative activity in computational shell mechanics. In the last three decades, numerous theoretical models have been developed and applied to various practical circumstances. It may be fair to state that no single theory has proven to be general and comprehensive enough for the entire range of applications".

In this book, the development of mixed formulation based, low-order, three-node flat triangular shell elements suitable for the linear and nonlinear analysis of thin to moderately thick shells is presented, together with their applications in the vibration characteristics and dynamic responses of complicated shell structures. It is the authors' hope that this book, in a very small way, continues the "innovative activities in computational shell mechanics".

Although as much details as needed regarding the development of the mixed formulation based three-node flat triangular shell elements are presented in two chapters, Chapters 2 and 7, this book is intended for those with some background in finite element analysis and numerical algorithms. Some familiarity with nonlinear mechanics is also assumed. As a result, the fundamental of finite element method is omitted. Readers may refer to [1.2, 1.3] for such topic.

1.2. ORGANIZATION OF BOOK

The book is organized into 9 chapters. Chapters 2 to 6 are concerned with the linear version of the mixed formulation based three-node flat triangular shell elements and their application in investigating the vibration characteristics of linear elastic structures. Specifically,

a) Chapter 2 presents the mixed formulation based three-node flat triangular shell elements within the context of linear analysis. It also examines issues such as rigid body modes, patch tests and mesh topologies;

b) Chapter 3 is concerned with the vibration analysis of plate structures;

c) Vibration characteristics of shells of single curvature and double curvatures are included in Chapters 4 and 5, respectively; and

d) Chapter 6 demonstrates the application of the shell elements to single-cell and double-cell box structures.

The remaining chapters, Chapters 7 to 9, deal with the general nonlinear dynamic analysis of shell structures by the mixed formulation based three-node flat triangular shell elements. The nonlinear formulation is given in Chapter 7, which is followed by Chapter 8 on the nonlinear dynamics of plate and box structures. Nonlinear dynamics of structures of single curvature and double curvatures are presented in Chapter 9. For the latter chapters, geometrical nonlinearity due to large deformation, material nonlinearity due to elastic-plastic material behaviour, and various loading situations including non-conservative pressure loads are investigated.

1.3. NOTES ON COMPUTER PROGRAMMING

The linear and nonlinear mixed formulation based three-node flat triangular shell elements were initially programmed in the Fortran language and incorporated in NONSAP [1.4] which was modified and implemented on a SGI workstation for the computational results reported in the doctoral degree thesis of the first author [1.5]. The digital computer program has since been rewritten in the personal computer (PC) based MATLAB system [1.6]. All computations involved in this book are performed in the MATLAB environment. Plots such as mode shapes and dynamic responses are generated by appropriate MATLAB functions.

At the present time, the shell element formulations and associated functions, such as mesh generation, applying boundary conditions, direct time integration schemes, Newton-Raphson method and its variants, Riks-Wempner arc-length method, plotting of mode shapes and time histories, and so on, are combined into a package, written by the first author for academic and research purpose only.

REFERENCES

[1.1]   H.T.Y. Yang, S. Saigal, A. Masud, and R. K. Kapania, "A survey of recent shell finite elements", *Int. J. Numer. Methods Eng.*, vol. 47, pp. 101-127, January 2000.

[1.2]   R.B. Cook, D.S. Malkus, and M.E. Plesha, *Concepts and Applications of Finite Element Analysis*, New York: John Wiley & Sons, 1989.

[1.3]   O.C. Zienkiewicz, and R.L. Taylor, *The Finite Element Method, I and II* (4th edn). New York: McGraw-Hill, 1991.

[1.4]   K.J. Bathe, E.L. Wilson, and R.H. Iding, "NONSAP, a structural analysis program for static and dynamic response of nonlinear systems", Report No. SESM 74-3, Structural Engineering Laboratory, University of California, Berkeley, California, 1974.

[1.5]   M.L. Liu, "Response statistics of shell structures with geometrical and material nonlinearities", Ph.D. thesis, The University of Western Ontario, London, ON, 1993.

[1.6]   The MathWorks Inc., *MATLAB The Language of Technical Computing*, Massachusetts, U.S.A.

6

Send Orders for Reprints to reprints@benthamscience.net
Vibration and Nonlinear Dynamics of Plates and Shells, 2014, 7-29

CHAPTER 2

Mixed Formulation Based Three-Node Flat Triangular Shell Elements for Vibration Analysis

Abstract: In order to investigate the vibration characteristics and dynamic responses of complicated shell structures with geometrical and material nonlinearities, it is essential to formulate shell finite elements that are easy to use, accurate, effective, and applicable to thin as well as moderately thick shells. This chapter presents the development of the mixed formulation or hybrid strain based three-node flat triangular shell elements, with a particular emphasis on the linear analysis of thin to moderately thick shells. Section 2.1 gives a brief introduction and an outline of the features of the shell elements. Section 2.2 deals with the derivation of consistent stiffness and mass matrices of a particular element. In Section 2.3, results and discussions pertaining to rigid-body modes, patch test, and mesh topology are presented. Concluding remarks are given in Section 2.4.

Keywords: Mixed formulation, three-nodes, triangular, shell finite elements, vibration analysis.

2.1. MIXED FORMULATION BASED THREE-NODE FLAT TRIANGULAR SHELL ELEMENTS

Low-order C^0 three-node flat triangular shell elements have attracted considerable attention since the early 1960s. In most of these elements, every node may have five or six degrees-of-freedom (DOF). Some notable early developments of elements with five nodal DOF include those by Zienkiewicz *et al.* [2.1], and Clough and Johnson [2.2], in which the five nodal DOF were, the two in-plane displacements, the transversal displacement and its two first derivatives with respect to the two axes perpendicular to the transversal displacement. On the other hand, Argyris *et al.* [2.3, 2.4] chose the mixed second-order derivative of the lateral displacement, together with the five DOF mentioned above, as nodal DOF, resulting in an element with a total of 18 DOF. Later in the 1980s, some three-node shell elements were developed [2.5, 2.6] by combining the three-node discrete Kirchhoff theory (DKT) triangular elements [2.7, 2.8] and the constant strain triangle (CST).

Another category of note-worthy shell elements are the so-called degenerate shell elements. They possess features such as mathematical consistency, easy extendability to nonlinear analysis and simplicity in formulation. However, two chief concerns exist. The first one is the shear-locking phenomenon. Degenerate elements perform satisfactorily with thick shells, but they become less accurate in the "thin" limit, owing to the excessive transversal shear strain involved in the formulation. Techniques have been proposed to reduce or even circumvent shear-locking. For a summary, readers are referred to Ref. [2.9]. One of the commonly adopted techniques is the reduced or selective integration [2.10]. Another scheme for dealing with the shear-locking problem is the hybrid/mixed formulation. The latter has been shown equivalent to the displacement-based formulation with reduced integration [2.9, 2.11], in terms of handling shear-locking. Furthermore, the hybrid/mixed formulation has the unique feature of providing continuity for the displacement field as well as the strain or stress field.

Degenerate shell elements may also exhibit membrane-locking when low-order in-plane displacement functions are used in the formulation of curved elements [2.12, 2.13]. Techniques to deal with membrane-locking include, for instance, utilization of enhanced membrane strain interpolations [2.14] and sufficiently high order in-plane displacement field [2.12]. It was also found that the reduced integration for shear strain also reduced the effects of membrane-locking [2.15]. It should be pointed out that for three-node degenerate shell elements specifically, the coupling between membrane and bending actions is unfortunately missing since the three nodes can only describe a "flat" geometry. However, owing to this "flat" geometry, membrane-locking is non-existent.

In addition, shell elements may encounter problems associated with the rotations about the normal to the shell surface, also known as the drilling degrees-of-freedom (DDOF). The DDOF are typically present among the structural or global DOF. If they are not included in shell elements that are coplanar at a certain node, the global stiffness matrix becomes singular. Omitting the rotations interferes with rigid body motion and thus destroys an important convergence criterion: correct representation of general rigid body motion. An approach in dealing with the DDOF was to include the normal rotation by employing curved membrane component element [2.16]. This shell element was later found to be identical to the Allman's triangle (AT) [2.17-2.19].

In summary, low-order, three-node flat triangular shell elements based on displacement formulation possess several important and advantageous features. From the modelling perspective, these flat elements model shell structures by the superposition of stretching behaviour (membrane element) and bending behaviour (plate bending element). Therefore, they are simple to formulate, easy to input data to describe general shell geometry, and able to represent rigid body motions. They can be mixed with other types of elements. That a relatively large number of elements may be required in a finite element model provides the advantage of convenience in incorporating complicated loading and boundary conditions. On the other hand, these low-order three-node flat triangular shell elements have shear-locking and problem caused by the DDOF. Consequently, it is necessary to develop three-node, 18-DOF, low-order flat triangular shell elements that have improved features over other flat triangular shell elements.

(i)   Such shell elements are degenerate in nature. The degenerate nature allows for applications to thin as well as moderately thick shells, and for relatively simple extension to nonlinear analysis.

(ii) Such shell elements are mixed formulation or hybrid strain based. The hybrid strain feature provides continuity in displacements and strains, and circumvents the shear-locking problem.

(iii) The choice of hybrid strain rather than hybrid stress is based on mathematical as well as practical rationale. Mathematically, variational principles are employed to minimize the strain energy in any element. As a result, strains converge more rapidly. Practically, the evaluation of element stiffness matrix involves relatively straightforward strain-displacement relationships (more precisely, incremental strain-displacement relationships for nonlinear problems), leading to more efficient computation. Conversely, stresses depend on materials (linear elastic or elasto-plastic in the nonlinear cases, for example) and the state of deformation.

(iv) A scheme similar to the Allman's triangle (AT) [2.17-2.19] is incorporated. As a result, in-plane displacements are coupled with the

DDOF, enabling the element to reflect the true normal rotations. In addition, incorporating the Allman's scheme improves the insufficiency of bending action which is typical when adopting low order interpolation. The resulting element is capable of representing true normal rotation as well as desirable membrane and bending behaviours.

(v) The element stiffness matrix can be and is explicitly expressed by a combination of manual and computer-assisted derivations. The explicit expressions eliminate the need for matrix inversion and numerical integration, and thereby improve the computational time.

In terms of vibration analysis of shell structures using the finite element method (FEM), one of the main considerations seems to be if and how to incorporate the rotary inertia, in addition to the translational inertia. One may argue for the practice to disregard the effects of rotary inertia due to bending when the shell is thin. However, when the shell structure is relatively thick the rotary inertia effect due to bending is not negligible. Moreover, it has been found that the DDOF play a crucial role in the performance of shell elements [2.20, 2.21]. Therefore, it is natural and necessary to bring in the effects due to torsional deformation associated with the DDOF.

Of special interest is Ref. [2.22], in which a set of explicit expressions for the mass and stiffness matrices of a triangular element was presented. The element mass matrix included translational inertia, and rotary inertia due to bending and torsion. To derive the element mass matrix, the triangle was divided into "beam elements" that were parallel to one of the three edges [2.22]. A relation between edge displacements and displacements within the triangle was established. This procedure was subsequently repeated on every side and the mean was applied to obtain the final element mass matrix, in order to eliminate the discrepancy due to forming "beam elements" parallel to one side or others. Consequently, the amount of computation involved was significant, not to mention that the physical interpretation of the element mass matrix was not apparent. In contrast, the present formulation is straightforward and has ample physical interpretation, as will be seen in Sub-sec. 2.2.9.

Additional features may be included in the formulation of the three-node flat triangular elements. These include, for example, linear distribution of the membrane part of the assumed strain field, and membrane-bending coupling by the hybrid strain formulation. After examining the effects of such additional features on the performance of the hybrid strain based three-node flat triangular shell elements, Ref. [2.23], however, found that the inclusion of such features was unnecessary. Instead, topology of the mesh played a crucial role.

2.2. THREE-NODE FLAT TRIANGULAR SHELL ELEMENTS

Various mixed formulation or hybrid strain based flat triangular shell finite elements have been previously developed and presented [2.20, 2.21]. However, in this book only two versions are employed for brevity and for their superior features over the other elements in [2.20-2.21].

The consistent element stiffness matrix k and consistent element mass matrix m of a mixed or hybrid-strain based three-node flat triangular shell element is presented in the following. The focus is on the shell element identified as AT+(k_t^1)' and AT+(k_t^3)' for static analysis [2.20], and as NFORMU = 15 and NFORM = 16 for vibration analysis [2.21]. These particular elements have been found to have superior performance to other hybrid-strain based three-node flat triangular shell elements developed at the same time [2.20-2.21, 2.24]. As will be observed in the following, these elements are mixed formulation based in the sense that the DDOF component of the element is derived from the displacement formulation while the remaining bending, membrane and shear components of the element are obtained through the hybrid strain formulation.

2.2.1. Variational Principle

Four major variational principles are acknowledged as the fundamentals of finite element formulations. They are: the principle of minimum potential energy, the principle of minimum complementary energy, the Hu-Washizu principle, and the Hellinger-Reissner principle. The last one assumes displacements and stresses, or displacements and strains, as independent variables, and is adopted in the present formulation for the reasons mentioned in Section 2.1. The functional for the

Hellinger-Reissner principle can be written as, when independent variables are displacements and strains:

$$\pi_{HR}(u,\varepsilon) = \int_{V_b}\left[-\frac{1}{2}\varepsilon^T\tilde{D}\varepsilon + \varepsilon^T\tilde{D}(Lu) - u^T f\right]dV - \int_{S_t}\left(u^T\bar{t}\right)dS - \int_{S_u}\left[(u-\bar{u})^T\Gamma(\tilde{D}\varepsilon)\right]dS \dots (2.1)$$

where σ is the stress vector, ε the strain vector, f the body force vector, $\tilde{D}$ the elastic matrix of the material (such that $\sigma = \tilde{D}\varepsilon$), L the linear operator to calculate strain from displacement, Γ the linear operator to evaluate surface traction from stress, $\bar{u}$ the vector of prescribed displacement on boundary, $\bar{t}$ the vector of prescribed surface traction, V_b the volume of the body, S_t the portion of the surface of the body where $\bar{t}$ is applied, and S_u the portion of the surface of the body where $\bar{u}$ is prescribed. Finally, the superscript "T" denotes transpose.

Since in hybrid strain formulation the final unknowns are nodal displacements, the satisfaction of displacement boundary condition, $u = \bar{u}$, is easily met. The term associated with $(u-\bar{u})$ can be disregarded. Equation (2.1) thus becomes,

$$\pi_{HR}(u,\varepsilon) = \int_{V_b}\left[-\frac{1}{2}\varepsilon^T D\varepsilon + \varepsilon^T D(Lu) - u^T f\right]dV - \int_{S_t}\left(u^T\bar{t}\right)dS \dots \tag{2.2}$$

2.2.2. Stationarity of Functional

Assuming that at the element level,

$$u = Nq \ , \ \varepsilon = \tilde{P}\beta \dots \tag{2.3 a,b}$$

where q and β are vectors of nodal displacements and strain parameters, respectively, while N and $\tilde{P}$ are the corresponding interpolation matrices. Substituting Eq. (2.3) into (2.2) leads to,

$$\pi_{HR}(u,\varepsilon) = \sum \pi_{HR}(q,\beta)$$

$$= \sum\left\{\int_{V_e}\left[-\tfrac{1}{2}\beta^T\tilde{P}^T\tilde{D}\tilde{P}\beta + \beta^T\tilde{P}^T\tilde{D}L(Nq)\right]dV\right\} - \sum\left\{\int_{V_e}\left(q^T N^T f\right)dV - \int_{A_e}\left(q^T N^T\bar{t}\right)dA\right\} \dots \tag{2.4}$$

where V_e and A_e are the volume and area of an element, respectively, and the summation is performed over all the elements.

Minimizing the total potential energy in Eq. (2.4) with respect to β, or setting $\dfrac{\partial \pi_{HR}(u, \beta)}{\partial \beta} = 0$ yields

$$-\left[\int_{V_e} \left(\tilde{P}^T \tilde{D} \tilde{P} \right) dV \right] \beta + \left[\int_{V_e} \left(\tilde{P}^T \tilde{D} B \right) dV \right] q = 0 \ldots \tag{2.5}$$

in which $B = LN$ is the linear strain-displacement matrix. Now defining

$$\tilde{H} = \int_{V_e} \left(\tilde{P}^T \tilde{D} \tilde{P} \right) dV \ , \quad G_e = \int_{V_e} \left(\tilde{P}^T \tilde{D} B \right) dV \ldots \tag{2.6 a,b}$$

such that β can be solved from Eq. (2.5)

$$\beta = \tilde{H}^{-1} G_e q \ldots \tag{2.7}$$

Substituting Eq. (2.7) into (2.4) results in

$$\pi_{HR}(u, \varepsilon) = \sum \pi_{HR}(q) = \sum \left[\tfrac{1}{2} q^T G_e^T \tilde{H}^{-1} G_e q \right] - \sum q^T \left[\int_{V_e} \left(N^T f \right) dV + \int_{A_e} \left(N^T \bar{t} \right) dA \right]$$

$$\ldots \tag{2.8}$$

Minimizing Eq. (2.8) with respect to q gives

$$G_e^T \tilde{H}^{-1} G_e q - \int_{V_e} \left(N^T f \right) dV - \int_{A_e} \left(N^T \bar{t} \right) dA = 0 \ldots \tag{2.9}$$

Finally, defining

$$k = G_e^T \tilde{H}^{-1} G_e \ , \quad f_c = \int_{V_e} \left(N^T f \right) dV + \int_{A_e} \left(N^T \bar{t} \right) dA \ldots \tag{2.10 a,b}$$

where k is the element stiffness matrix and f_e the consistent load vector due to body force and surface traction, Eq. (2.9) can then be simplified

$$kq = f_e \ldots \tag{2.11}$$

Assembling all elements yields the equilibrium equation of the structure

$$KQ = F_e \ldots \tag{2.12}$$

with K, Q and F_e being the assembled stiffness matrix, assembled nodal displacement vector and assembled consistent load vector, respectively. The unknowns in Eq. (2.12) are vector Q which, after being solved, is applied to recover the strain or stress vector by the following relations

$$\varepsilon = \tilde{P}\beta = \tilde{P}\tilde{H}^{-1}G_e q \,, \quad \sigma = \tilde{D}\varepsilon = \tilde{D}\tilde{P}\tilde{H}^{-1}G_e q \ldots \tag{2.13 a,b}$$

2.2.3. Geometry and DOF of Element

The geometry of the three-node flat triangular shell element is shown in Fig. **2.1**. The three nodes are allocated at the three corners of the mid-surface of the element. A local Cartesian coordinate system *r-s-t* is attached to Node 1, where the *r*-axis coincides with side 1-2; the *t*-axis is parallel to the normal to the plane containing the element, while the *s*-axis is perpendicular to the plane. As a result, the local coordinates of the three nodes are, $(0, 0)$, (r_2, s_2) and $(r_3, 0)$, respectively. The six local DOF associated with each node are the displacements u, v and w in the r, s and t directions, respectively, and the rotations θ_r, θ_s and θ_t about the r, s, and t-axes, respectively. The last rotation is also known as the DDOF, the drilling degree-of-freedom. The displacements are considered positive if along the positive directions of r, s, and t-axes. For the rotations, the right-hand screw rule is adopted in determining their directions.

The triangular element can be easily described by the natural or area co-ordinate system. If ξ_i ($i = 1$, 2 and 3) are the natural coordinates, they satisfy the condition

$$\xi_1 + \xi_2 + \xi_3 = 1 \ldots \tag{2.14}$$

The relation between the natural and the *r-s* coordinates is

$$\left\{ \begin{array}{c} \xi_1 \\ \xi_2 \\ \xi_3 \end{array} \right\} = \frac{1}{r_2 s_3} \begin{bmatrix} r_2 s_3 & -s_3 & r_3 - r_2 \\ 0 & s_3 & -r_3 \\ 0 & 0 & r_2 \end{bmatrix} \left\{ \begin{array}{c} 1 \\ r \\ s \end{array} \right\} \ldots \tag{2.15}$$

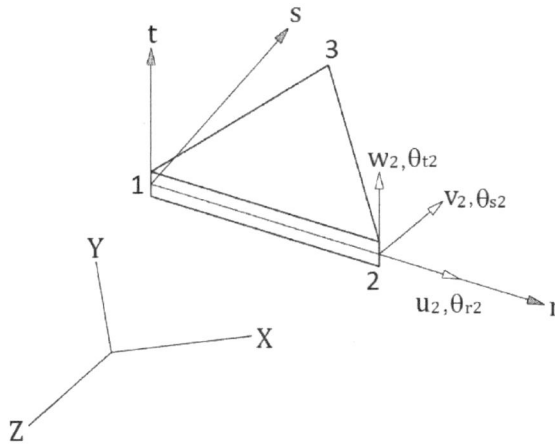

Figure 2.1: Nodes, axes and DOF of three-node flat triangular shell element.

The first-order partial derivatives of ξ_i (i = 1, 2 and 3) with respect to r and s are

$$\xi_{1,r} = -\frac{1}{r_2} \ , \ \xi_{1,s} = \frac{r_3 - r_2}{r_2 s_3} \ , \ \xi_{2,r} = \frac{1}{r_2} \ , \ \xi_{2,s} = -\frac{r_3}{r_2 s_3} \ , \ \xi_{3,r} = 0 \ , \ \xi_{3,s} = \frac{1}{s_3} \ldots \tag{2.16}$$

where the comma partial derivative notation in the subscripts has been adopted.

2.2.4. Assumed Displacement Field within Element

The assumed displacement field within an element is

$$\begin{bmatrix} u & v & w & \theta_r & \theta_s & \theta_t \end{bmatrix}^T = N \begin{bmatrix} u_1 & v_1 & w_1 & \theta_{r1} & \theta_{s2} & \theta_{t2} & u_2 & v_2 & \ldots\ldots & \theta_{t3} \end{bmatrix}^T \ldots \tag{2.17}$$

where the displacement interpolation or shape function matrix N is given by

$$N = \begin{bmatrix} N_1 & N_2 & N_3 \end{bmatrix} \ldots \tag{2.18}$$

with the sub-matrices N_i (i = 1, 2 and 3) being defined as

$$N_i = \begin{bmatrix} \xi_i & & & & & \bar{p}_i \\ & \xi_i & & & & \bar{q}_i \\ & & \xi_i & -\bar{p}_i & -\bar{q}_i & \\ & & & \xi_i & & \\ & & & & \xi_i & \\ & & & & & \xi_i \end{bmatrix} \ldots \tag{2.19}$$

It should be noted that, for brevity, only the non-zero elements of the sub-matrices are given in Eq. (2.19). Although this will remain for the remainder of the book, zeros will sometimes be filled in for easy reference. The terms $\bar{p}_i$ and $\bar{q}_i$ are,

$$
\begin{aligned}
\bar{p}_1 &= \left(a_{31}\xi_3 - a_{12}\xi_2\right)\xi_1 & \bar{q}_1 &= \left(b_{31}\xi_3 - b_{12}\xi_2\right)\xi_1 \\
\bar{p}_2 &= \left(a_{12}\xi_1 - a_{23}\xi_3\right)\xi_2 & \bar{q}_2 &= \left(b_{12}\xi_1 - b_{23}\xi_3\right)\xi_2 \quad \ldots \\
\bar{p}_3 &= \left(a_{23}\xi_2 - a_{31}\xi_1\right)\xi_3 & \bar{q}_3 &= \left(b_{23}\xi_2 - b_{31}\xi_1\right)\xi_3
\end{aligned}
\tag{2.20}
$$

where the quantities a_{ij} and b_{ij} are, with reference to Fig. **2.2**,

$$
\begin{aligned}
a_{12} &= \tfrac{1}{2}\ell_{12}\cos\gamma_{12} & b_{12} &= \tfrac{1}{2}\ell_{12}\sin\gamma_{12} \\
a_{23} &= \tfrac{1}{2}\ell_{23}\cos\gamma_{23} & b_{23} &= \tfrac{1}{2}\ell_{23}\sin\gamma_{23} \quad \ldots \\
a_{31} &= \tfrac{1}{2}\ell_{31}\cos\gamma_{31} & b_{31} &= \tfrac{1}{2}\ell_{31}\sin\gamma_{31}
\end{aligned}
\tag{2.21}
$$

It should be mentioned that, (i) the $\bar{p}_i$ and $\bar{q}_i$ terms in the first two rows of the N matrix couple the in-plane displacements u and v with the DDOF, ensuring that the element will reflect the true normal rotations; (ii) the lateral displacement w is linear with regard to nodal lateral displacements w_i ($i = 1, 2, 3$), but is quadratic in terms of nodal rotations θ_{ri} and θ_{si} ($i = 1, 2, 3$), because of the $\bar{p}_i$ and $\bar{q}_i$ terms in the third row of the N matrix. As pointed out by [2.25], adding the $\bar{p}_i$ and $\bar{q}_i$ terms makes the element softer, improving the insufficiency of bending action due to the use of low order interpolation.

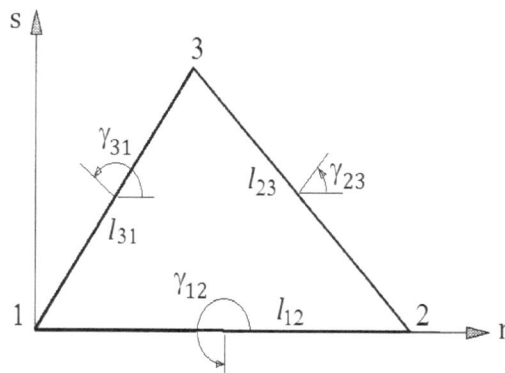

Figure 2.2: Geometry of three-node flat triangular shell element.

2.2.5. Strain Field within Element

The assumed strain field is

$$\begin{bmatrix} \varepsilon_r & \varepsilon_s & \varepsilon_{rs} & \varepsilon_{st} & \varepsilon_{tr} \end{bmatrix}^T = \tilde{P} \begin{bmatrix} \beta_1 & \beta_2 & \beta_3 & \ldots\ldots & \beta_8 & \beta_9 \end{bmatrix}^T \ldots \tag{2.22}$$

In Eq. (2.22), the β_i ($i = 1,..,9$) are strain parameters. Matrix $\tilde{P}$ is,

$$\tilde{P} = \begin{bmatrix} 1 & & & t & & \\ & 1 & & & t & \\ & & 1 & & & t \\ & & & -s_3(1-2\xi_2) & s_3(2\xi_2+2\xi_3-1) & \\ & & & -r_3(1-2\xi_2) & r_{32}(2\xi_2+2\xi_3-1) & r_2(1-2\xi_3) \end{bmatrix} \ldots \tag{2.23}$$

where $r_{32} = r_3 - r_2$, and t is the thickness coordinate whose value ranges $-h/2$ to $h/2$ with h being the thickness of the shell element. The $\tilde{P}$ matrix ensures that the nine strain parameters are evenly distributed over the membrane, bending and transverse shear strain fields. Specifically, β_1 through β_3 describe a constant membrane strain field, β_4 through β_6 correspond to a constant bending curvature distribution, and β_7 through β_9 represent a constant (over the thickness) transverse shear strain field.

2.2.6. Constitutive Relations

The constitutive relations for a homogeneous, isotropic and linearly elastic material, or $\sigma = \tilde{D}\,\varepsilon$, are, when written in details:

$$\begin{Bmatrix} \sigma_r \\ \sigma_s \\ \sigma_{rs} \\ \sigma_{st} \\ \sigma_{tr} \end{Bmatrix} = \begin{bmatrix} \dfrac{E}{1-v^2} & \dfrac{vE}{1-v^2} & & & \\ \dfrac{vE}{1-v^2} & \dfrac{E}{1-v^2} & & & \\ & & G & & \\ & & & \kappa_s G & \\ & & & & \kappa_s G \end{bmatrix} \begin{Bmatrix} \varepsilon_r \\ \varepsilon_s \\ \varepsilon_{rs} \\ \varepsilon_{st} \\ \varepsilon_{tr} \end{Bmatrix} \ldots \tag{2.24}$$

where E is the Young's modulus, G the shear modulus, v the Poisson's ratio and κ_s = 5/6 the form factor of shear. Note that the shear strains ε_{rs}, ε_{st} and ε_{tr} are the so-called engineering strains.

2.2.7. Element Stiffness Matrix k_H

The consistent element stiffness matrix k will be presented in two sub-sections. This sub-section is concerned with the first part of the matrix, denoted as k_H which is determined per Eq. (2.10a) or per hybrid strain formulation. It is found that k_H can be divided into component matrices that represent the effects due to membrane, bending, and transversal shear, indicated by the subscripts m, b and s, respectively, as follows:

$$k_H = k_m + k_b + k_s \ldots \tag{2.25}$$

where

$$k_m = \left(G_e\right)_m^T \tilde{H}^{-1}\left(G_e\right)_m , \; k_b = \left(G_e\right)_b^T \tilde{H}^{-1}\left(G_e\right)_b , \; k_s = \left(G_e\right)_s^T \tilde{H}^{-1}\left(G_e\right)_s \ldots \tag{2.26 a-c}$$

The details of the matrices such as $(G_e)_m$ and so on, are,

$$\left(G_e\right)_m = \int_{V_e}\left(\tilde{P}^T \tilde{D}B_m\right)dV , \; \left(G_e\right)_b = \int_{V_e}\left(\tilde{P}^T \tilde{D}B_b\right)dV , \; \left(G_e\right)_s = \int_{V_e}\left(\tilde{P}^T \tilde{D}B_s\right)dV \ldots \tag{2.27 a-c}$$

where B_m, B_b and B_s are 5×18 matrices. They can be written as follows,

$$B_m = \begin{bmatrix} B_{m1} & B_{m2} & B_{m3} \end{bmatrix} , \; B_b = \begin{bmatrix} B_{b1} & B_{b2} & B_{b3} \end{bmatrix} , \; B_s = \begin{bmatrix} B_{s1} & B_{s2} & B_{s3} \end{bmatrix} \ldots \tag{2.28 a-c}$$

Details of the 5×6 sub-matrices B_{mi}, B_{bi} and B_{si} (i = 1, 2 and 3) are

$$B_{mi} = \begin{bmatrix} \xi_{i,r} & & 0 & 0 & 0 & \overline{p}_{i,r} \\ & \xi_{i,s} & & & & \overline{q}_{i,s} \\ \xi_{i,s} & \xi_{i,r} & & & & \overline{p}_{i,s} + \overline{q}_{i,r} \\ & & & & & \\ & & & & & \\ & & & & & \end{bmatrix} , \; B_{bi} = \begin{bmatrix} 0 & 0 & 0 & & t\xi_{i,r} & 0 \\ & & & -t\xi_{i,s} & & \\ & & & -t\xi_{i,r} & t\xi_{i,s} & \\ & & & & & \\ & & & & & \\ & & & & & \end{bmatrix} \ldots \tag{2.29 a-b}$$

$$
B_{si} = \begin{bmatrix}
 & & & & & & \cdots \\
 & & \xi_{i,s} & -\overline{P}_{i,s} & \xi_i - \overline{q}_{i,s} & \\
0 & 0 & \xi_{i,r} & -\xi_i - \overline{P}_{i,r} & -\overline{q}_{i,r} & 0
\end{bmatrix}
\qquad (2.29\ c)
$$

2.2.8. Consistent Element Stiffness Matrix *k*

The matrix k_H, as defined in Eq. (2.25), does not include the (torsional) effect due to DDOF. Computationally, this will cause singularity or rank deficiency. A notable attempt in dealing with the DDOF was proposed by Kanok-Nukulchai [2.26], in which a penalty function was introduced to complement the strain energy functional. That is

$$
\pi(q,\beta) = \pi_{HB}(q,\beta) + \pi_t(q) \ldots
\qquad (2.30)
$$

with

$$
\pi_t = \kappa G h \sum \left\{ \int_{A_e} \left[\theta_t - \tfrac{1}{2}(v_{,r} - u_{,s}) \right]^2 dA \right\} \ldots
\qquad (2.31)
$$

Since $\tfrac{1}{2}(v_{,r} - u_{,s})$ is the (averaged) normal rotation from elasticity theory, Eq. (2.31) forces θ_t to approach the "true" normal rotation, resulting in the desired constraint. Clearly, if $\kappa = 1/2$, the π_t of Eq. (2.31) becomes the strain energy due to torsional deformation. By setting $\kappa = 1/2$ and substituting Eq. (2.17) into (2.31), the latter becomes

$$
\pi_t = q^T \left[\frac{1}{2} G h \int_{A_e} \left(\overline{Y}^T \overline{Y} \right) dA \right] q \ldots
\qquad (2.32)
$$

In other words, the consistent element stiffness matrix is

$$
k = k_H + k_t \ldots
\qquad (2.33)
$$

with

$$k_t = \frac{1}{2} Gh \int_{A_e} \left(\overline{Y}^T \overline{Y} \right) dA \dots \tag{2.34}$$

and

$$\overline{Y} = \frac{1}{2} \left[\frac{r_3 - r_2}{r_2 s_3}, \quad \frac{1}{r_2}, \quad 0, \quad 0, \quad 0, \quad 2\xi_1 - \overline{q}_{1,s} + \overline{p}_{1,r}, \quad -\frac{r_3}{r_2 s_3}, \quad -\frac{1}{r_2}, \quad 0, \quad 0, \right.$$
$$\left. \quad 0, \quad 2\xi_2 - \overline{q}_{2,s} + \overline{p}_{2,r}, \quad \frac{1}{s_3}, \quad 0, \quad 0, \quad 0, \quad 0, \quad 2\xi_3 - \overline{q}_{3,s} + \overline{p}_{3,r} \right] \tag{2.35}$$

In Ref. [2.26] it was suggested to use one-point quadrature to evaluate Eq. (2.34) so as to avoid an over-constrained situation similar to shear-locking. This one-point quadrature can be easily achieved by setting $\xi_1 = \xi_2 = \xi_3 = 1/3$ in Eq. (2.34). The resulting k_t together with the k_H of Sub-sec. 2.2.7 form the stiffness matrix labelled as AT+(k_t^1)' in [2.20]. For three-point quadrature, or full integration of Eq. (2.34), the resulting consistent element matrix is identified as AT+(k_t^3)'. References [2.20, 2.24] showed that using one-point quadrature in fact yielded better results than using the full integration. However, the AT+(k_t^3)' element is able to provide the correct number of rigid body modes while AT+(k_t^1)' has two spurious modes [2.21].

2.2.9. Element Consistent Mass Matrix *m*

The usual definition of the consistent mass matrix of an element is

$$m = \int_{V_e} \rho_0 \left(N^T N \right) dV \dots \tag{2.36}$$

where ρ_0 is the mass density of the material. This definition, however, is only applicable to translational DOF such as *u*, *v* and *w*. For rotational DOF, Eq. (2.36) leads to results that have no physical meaning. This may be one of the reasons that rotational DOF are sometimes disregarded in free vibration analysis by the FEM [2.27]. In [2.28] it was suggested to replace ρ_0 by a diagonal matrix $\tilde{\rho}$ whose elements are, ρ_0 for translational DOF; and $\rho_0 h^2/12$, which is the moment of inertia (per unit cross-sectional width), for rotational DOF pertaining to bending. However, no detail on the elements in the mass matrix associated with DDOF was given in [2.28]. In the mixed formulation based three-node flat triangular element, the elements in $\tilde{\rho}$ associated with the DDOF are chosen to be the polar moment of inertia of the triangle (per unit area of the triangle) about its centroid. That is, the elements in $\tilde{\rho}$ associated with the DDOF are

$$J_d = \rho_0 \frac{r_2^2 + s_3^2 + r_3\left(r_3 - r_2\right)}{18} \ldots \tag{2.37}$$

Now, the definition of consistent element mass matrix becomes

$$m = \int_{V_e}\left(N^T \tilde{\rho} N\right) dV \ldots \tag{2.38}$$

where

$$\tilde{\rho} = \begin{bmatrix} \rho_0 & & & & & \\ & \rho_0 & & & & \\ & & \rho_0 & & & \\ & & & \dfrac{\rho_0 h^2}{12} & & \\ & & & & \dfrac{\rho_0 h^2}{12} & \\ & & & & & J_d \end{bmatrix} \ldots \tag{2.39}$$

Finally, it should be emphasized that the integrations per Eqs. (2.6a), (2.10b), (2.27) and (2.38) have been performed analytically with the aid of a symbolic algebraic manipulation package, MAPLE [2.29]. It is "exact" in the sense that no numerical integration is used. A more precise description of these "exact" stiffness and mass matrices is referred to as explicit expressions of stiffness and mass matrices.

2.3. RIGID-BODY MODES, PATCH TEST AND MESH TOPOLOGY

Before the mixed formulation based three-node flat triangular shell elements, as presented in the previous section, are to be used for the vibration analysis of linear plate, shell and celled structures in the following chapters, the two elements $AT+(k_t^1)$' and $AT+(k_t^3)$' are applied for the rigid body mode tests, and for the patch test. Passing both tests will ensure convergence of finite element solutions toward the correct one.

2.3.1. Rigid-Body Modes

As reported in [2.21] unconstrained (free-free) single element tests indicated that the full integration version $AT+(k_t^3)$' can give the six rigid body modes correctly. However, the one point quadrature $AT+(k_t^1)$' results in two spurious modes.

2.3.2. Patch Test

A patch test is to verify that a group (or a patch) of arbitrarily oriented elements is able to represent a constant strain field. Passing the patch test is the necessary and sufficient condition of convergence, as concluded in Ref. [2.30].

The patch test performed here is described in Appendix B of [2.31] together with its analytical solutions (see Fig. **2.3** and Tables **2.1** through **2.3**). It involves a rectangular plate which is divided into four triangular elements. The geometrical dimensions of the plate are $1 \times 2 \times 0.001$ m^3. Its material properties are: Young's modulus $E = 10^8$ Pa and Poisson's ratio $v = 0.25$. The fifth node has the co-ordinates of $X_5 = 0.7$ m and $Y_5 = 0.9$ m, although other values are also used. By prescribing boundary conditions and bending moments in accordance with Tables **2.1** through **2.3**, states of constant bending curvature and constant twist are reached. The analytical solutions and computed finite element results are presented in Table **2.4**. It is evident that the triangular element passes the patch test. In fact, other thicknesses such as 0.0001 m and 0.1 m have also been tried, and the same degree of agreement with the analytical solutions as in Table **2.4** has been observed. In passing, it is noted that in performing the patch test, nodal DOF associated with displacements in the X and Y directions and rotations about the Z axis are all constrained. In Tables **2.1** through **2.3**, m_1 and m_2 are applied end moments.

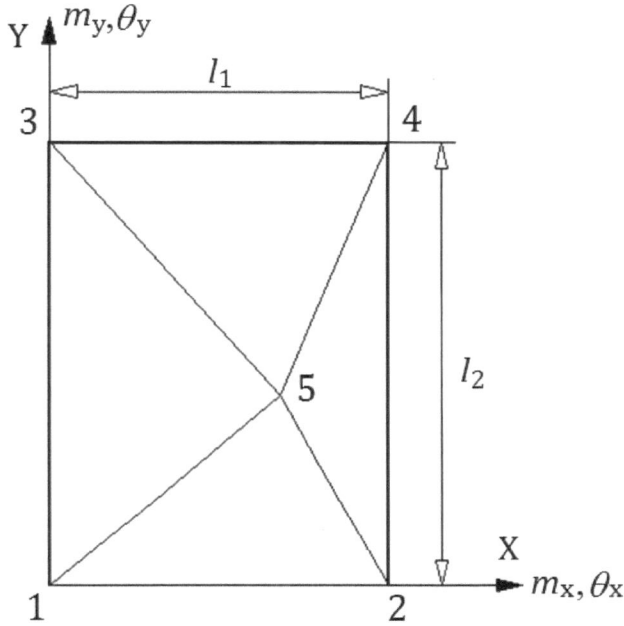

Figure 2.3: Four triangular elements used in patch test.

Table 2.1: Patch test of constant curvature bending (*X*-direction).

Description of Boundary Conditions and Applied Loads						
Node	Boundary Conditions			Applied Loads		
	W	Θ_X	Θ_Y	f_z	m_x	m_y
1	fixed	fixed	free	0	0	$-m_2$
2	fixed	fixed	free	0	0	m_2
3	fixed	fixed	free	0	0	$-m_2$
4	fixed	fixed	free	0	0	m_2
5	free	fixed	free	0	0	0
Analytical Solution for Displacement						
Node	W	Θ_X	Θ_Y			
1	0	0	$-\theta$	$\theta = \dfrac{m_2 l_1}{D l_2}$ $D = \dfrac{Eh^3}{12(1-v^2)}$		
2	0	0	θ			
3	0	0	$-\theta$	$a = x_5\left(1-\dfrac{x_5}{l_1}\right)$ $b = \dfrac{2x_5}{l_1}-1$		
4	0	0	θ			
5	$a\theta$	0	$b\theta$			

Table 2.2: Patch test of constant curvature bending (*Y*-direction).

Description of Boundary Conditions and Applied Loads						
Node	Boundary Conditions			Applied Loads		
	W	Θ_X	Θ_Y	f_z	m_x	m_y
1	fixed	free	fixed	0	m_1	0
2	fixed	free	fixed	0	m_1	0
3	fixed	free	fixed	0	$-m_1$	0
4	fixed	free	fixed	0	$-m_1$	0
5	free	free	fixed	0	0	0
Analytical Solution for Displacement						
Node	W	Θ_X	Θ_Y			
1	0	θ	0			
2	0	θ	0			
3	0	$-\theta$	0			
4	0	$-\theta$	0			
5	$a\theta$	$b\theta$	0			

$$\theta = \frac{m_1 l_2}{D l_1} \quad D = \frac{Eh^3}{12\left(1-v^2\right)}$$

$$a = y_5\left(1 - \frac{y_5}{l_2}\right) \quad b = 1 - \frac{2y_5}{l_2}$$

Table 2.3: Patch test of constant twist.

Description of Boundary Conditions and Applied Loads						
Node	Boundary Conditions			Applied Loads		
	W	Θ_X	Θ_Y	f_z	m_x	m_y*
1	fixed	free	free	0	$-m_1$	m_2
2	fixed	free	free	0	m_1	m_2
3	fixed	free	free	0	$-m_1$	$-m_2$
4	free	free	free	0	m_1	$-m_2$
5	free	free	free	0	0	0
Analytical Solution for Displacement						
Node	W	Θ_X	Θ_Y			
1	0	0	0			
2	0	κl_1	0			
3	0	0	$-\kappa l_2$			
4	$\kappa l_1 l_2$	κl_1	$-\kappa l_2$			
5	$\kappa x_5 y_5$	κx_5	$-\kappa y_5$			

$$m = \frac{2m_1}{l_2} \quad \kappa = \frac{m}{D\left(1-v\right)}$$

$$D = \frac{Eh^3}{12\left(1-v^2\right)}$$

*$m_2 = m_1 l_1 / l_2$.

Table 2.4: Finite element results of patch test for bending.

Node	Analytical			Triangular Element		
Constant Bending Curvature in X-Direction, m_2=1.0						
	W	Θ_X	Θ_X	W	Θ_X	Θ_Y
1	0	0	-56.25	0	0	-56.25
2	0	0	56.25	0	0	56.25
3	0	0	-56.25	0	0	-56.25
4	0	0	56.25	0	0	56.25
5	11.8125	0	22.50	11.812	0	22.50
Constant Bending Curvature in Y-Direction, m_1=1.0						
	W	Θ_X	Θ_Y	W	Θ_X	Θ_Y
1	0	225.0	0	0	225.0	0
2	0	225.0	0	0	225.0	0
3	0	-225.0	0	0	-225.0	0
4	0	-225.0	0	0	-225.0	0
5	111.375	22.50	0	111.37	22.50	0
Constant Twist, m_1=1.0, m_2=0.5						
	W	Θ_X	Θ_Y	W	Θ_X	Θ_Y
1	0	0	0	0	α^*	$-\beta^*$
2	0	150.0	0	0	150.0	β^*
3	0	0	-300.0	0	$-\alpha^*$	-300.0
4	300.0	150.0	-300.0	300.0	150.0	-300.0
5	94.50	105.0	-135.0	94.50	105.0	-135.0

* $\alpha = 9.9509 \times 10^{-9}$, $\beta = 1.0709 \times 10^{-8}$.

2.3.3. Mesh Topology

When the mid-surface of the plate or shell is of a quadrilateral (straight or curved sided) shape, four types of mesh can be formed. They are shown in Fig. **2.4** and referred to as types A, B, C or D. Reference [2.23] observed that results using mesh types C and D had little to no difference, and that mesh type C or D was superior to type A or B in terms of accuracy, given the same effective DOF or the number of total unknown displacements.

On the other hand, when the mid-surface of the plate or shell is three-sided, straight or curved, the two mesh types given in Fig. **2.5** may be considered. The mesh type on the left has equal number of nodes on the various radii. Mesh type A, B, C or D is formed from the second radial division and outward. In [2.20, 2.24] this type of mesh was referred to by the number of elements and the type letter A, B, C or D. On the other hand, the mesh type on the right increases the number of nodes by two and the number of elements by four, each time an outward radial division is added. This mesh type will be identified by the number of elements and letter F. The comparison of such meshes is given in the coming chapters.

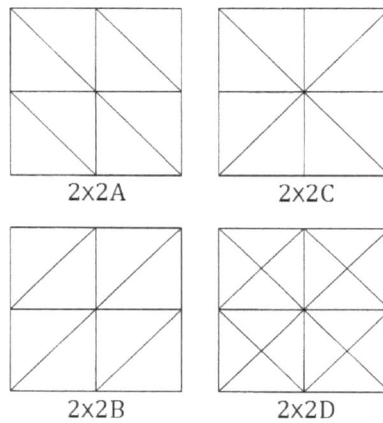

Figure 2.4: Mesh types A, B, C and D.

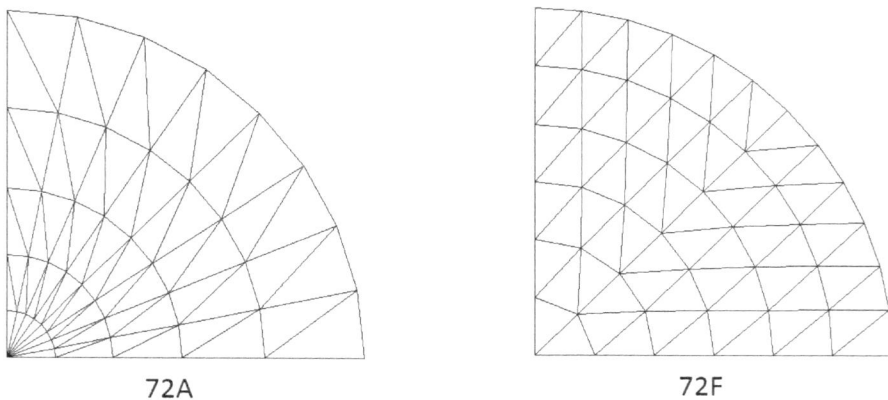

Figure 2.5: Mesh types used with three-sided area or surface.

2.4. CONCLUDING REMARKS

In this chapter the important and advantageous features of low-order, and three-node flat triangular shell elements have been presented. The need for developing such elements in a way that they are of the mixed formulation type and degenerate in nature has been addressed. Details of the consistent element stiffness matrix *k* and mass matrix *m* were presented for two particular versions of the shell element, $AT+(k_t^1)$' and $AT+(k_t^3)$'. The two elements have been tested for rigid-body modes and patches. Both elements were found to pass the patch test. Thus, convergence toward the correct result is assured. Mesh types used with quadrilateral or triangular domain, either straight or curved sided, were examined. The mixed formulation based three-node flat triangular shell elements have been applied to the vibration analysis of linear plate, shell and celled structures, the results of which are included in the next four chapters.

DISCLOSURE

Part of the presentation included in this chapter has been previously published in Finite Elements in Analysis and Design, Volume 17, Issue 3, October 1994, Pages 169 – 203, and in Journal of Sound and Vibration, Volume 184, Issue 5, August 1995, Pages 801 – 821.

REFERENCES

[2.1] O.C. Zienkiewicz, C.J. Parikh, and I.P. King, "Arch dam analysis by a linear finite element shell solution program", in *Proc. Symposium on Arch Dams*, 1968, pp. 19-22.

[2.2] R.W. Clough, and C.P. Johnson, "A finite element approximation for the analysis of thin shell", *Int. J. Solids Struct.*, vol. 4, pp. 43-60, January 1968.

[2.3] J.H. Argyris, P.C. Dunne, G.A. Malejannakis, and E. Schelke, "A simple triangular facet shell element with applications to linear and nonlinear equilibrium and elastic stability problems", *Comp. Methods Appl. Mech. Eng.*, vol. 10, pp. 371-403, March 1977.

[2.4] J.H. Argyris, P.C. Dunne, G.A. Malejannakis, and E. Schelke, "A simple triangular facet shell element with applications to linear and nonlinear equilibrium and elastic stability problems", *Comp. Methods Appl. Mech. Eng.*, vol. 11, pp. 97-131, April 1977.

[2.5] K.J. Bathe, and L.W. Ho, "A simple and effective element for analysis of general shell structures", *Comp. Struct.*, vol. 13, pp. 673-681, October 1981.

[2.6] J.L. Batoz, and G. Dhatt, "Development of two simple shell elements", *AIAA J.*, vol. 10, pp. 237-248, February 1972.

[2.7] J.L. Batoz, K.J. Bathe, and L.W. Ho, "A study of three-node triangular plate bending elements", *Int. J. Numer. Methods Eng.*, vol. 15, pp. 1771-1812, December 1980.

[2.8] J.L. Batoz, "An explicit formulation for an efficient triangular plate-bending element", *Int. J. Numer. Methods Eng.*, vol. 18, pp. 1077-1089, July 1982.

[2.9] H.T.Y. Yang, S. Saigal, and D.G. Liaw, "Advances of thin shell finite elements and some applications, Version I", *Comp. Struct.*, vol. 35, pp. 481-504, January 1990.

[2.10] O.C. Zienkiewicz, R.L. Taylor, and J.M. Too, "Reduced integration technique in general analysis of plates and shells", *Int. J. Numer. Methods Eng.*, vol. 3, pp. 275-290, April/June 1971.

[2.11] D.G. Kang, "Present finite element technology from a hybrid formulation perspective", *Comp. Struct.*, vol. 35, pp. 321-329, January 1990.

[2.12] H. Stolarski, and T. Belystchko, "Membrane locking and reduced integration for curved elements", *J. Appl. Mech. ASME*, vol. 49, pp. 172-176, March 1982.

[2.13] H. Parisch, "A critical survey of the 9-node degenerated shell element with special emphasis on thin shell element application and reduced integration", *Comp. Methods Appl. Mech. Eng.*, vol. 20, pp. 323-350, December 1979.

[2.14] H.C. Huang, and E. Hinton, "A new nine node degenerated shell element with enhanced membrane and shear interpolation", *Int. J. Numer. Methods Eng.*, vol. 22, pp. 73-92, January 1986.

[2.15] H. Stolarski, and T. Belystchko, "Shear and membrane locking in curved C^0 elements", *Comp. Methods Appl. Mech. Eng.*, vol. 41, pp. 279-296, December 1983.

[2.16] N. Carpenter, H. Stolarski, and T. Belystchko, "A flat triangular shell element with improved membrane interpolation", *Commun.Appl. Numer. Methods*, vol. 1, pp. 161-168, July 1985.

[2.17] D.J. Allman, "A compatible triangular element including vertex rotations for plane elasticity analysis", *Comp. Struct.*, vol. 19, pp. 1-8, January1984.

[2.18] D.J. Allman, "The constant strain triangle with drilling rotations: A simple prospect for shell analysis", in *Proceedings the Mathematics of Finite Elements and Applications*, 1987, pp. 230-236.

[2.19] D.J. Allman, "Evaluation of the constant strain triangle with drilling rotations", *Int. J. Numer. Methods Eng.*, vol. 26, pp. 2645-2655, December 1988.

[2.20] C.W.S. To, and M.L. Liu, "Hybrid strain based three-node flat triangular shell elements", *Finite Elements in Analysis and Design*, vol. 17, pp. 169-203, October 1994.

[2.21] M.L. Liu, and C.W.S. To, "Vibration of structures by hybrid strain based three-node flat triangular shell elements", *J. Sound Vibr.*, vol. 184, pp. 801-821, August 1995.

[2.22] R. Dungar, R.T. Severn, and P.R. Taylor, "Vibration of plate and shell structures using triangular finite elements", *J. Strain Analysis*, vol. 2, pp. 73-83, January 1967.

[2.23] M.L. Liu, and C.W.S. To, "A further study of hybrid strain-based three-node flat triangular shell elements", *Finite Elements in Analysis and Design*, vol. 31, pp. 135-152, December 1998.

[2.24] M.L. Liu, "Response statistics of shell structures with geometrical and material nonlinearities," Ph.D. thesis, The University of Western Ontario, London, ON, 1993.

[2.25] A. Tessler, and T.J.R. Hughes, "A three-node Mindlin plate element with improved transverse shear", *Comp. Methods Appl. Mech. Eng.*, vol. 50, pp. 71-101, July 1985.

[2.26] W. Kanok-Nukulchai, "A simple and efficient finite element for general shell analysis", *Int. J. Numer. Methods Eng.*, vol.14, pp. 179-200, 1979.

[2.27] R.H. MacNeal, "A simple quadrilateral shell element", *Comp. Struct.*, vol. 8, pp. 175-183, April 1978.

[2.28] O.C. Zienkiewicz, and R.L. Taylor, *The Finite Element Method, I and II* (4th edn). New York: McGraw-Hill, 1991.

[2.29] Symbolic Computation Group, *MAPLE Reference Manual* (5th ed.), Department of Computer Science, University of Waterloo, Canada (1988).

[2.30] R. L. Taylor, J. C. Simo, O. C. Zienkiewicz, and A. C. H. Chan, "The patch test – a condition for assessing FEM convergence", *Int. J. Numer. Methods Eng.*, vol.22, pp. 39-62, January 1986.

[2.31] N. Carpenter, T. Belystchko, and H. Stolarski, "Locking and shear scaling factors in C^0 bending elements", *Comp. Struct.*, vol. 22, pp. 39-52, February 1986.

Send Orders for Reprints to reprints@benthamscience.net

CHAPTER 3

Vibration Analysis of Plate Structures

Abstract: This chapter is concerned with the vibration analysis of linear plates employing the mixed formulation based three-node flat triangular shell elements in Chapter 2. Square, circular and skew plates are studied. In particular, circular plates with various aspect ratios and boundary conditions are included. As a special example to reveal the features of the membrane component of the shell elements a plane stress problem is presented.

Keywords: Vibration analysis, plates, square, circular, skew.

3.1. SQUARE PLATE WITH VARIOUS BOUNDARY CONDITIONS

The geometrical dimensions of the square plate are $1 \times 1 \times 0.005$ m^3. Its material properties are: Young's modulus $E = 200$ GPa, Poisson's ratio $v = 0.3$ and mass density $\rho_0 = 7830$ kg/m^3. For finite element representation, mesh types C and D are employed to discretize the entire plate. Owing to the nature of the problem, the two in-plane displacements, U and V, and the rotation about Z-axis, Θ_Z, are constrained to zero. Note that capital alphabets indicate displacements or rotations in the global coordinates, unless stated otherwise. A representative 2×2C mesh of the plate is given in Fig. **3.1**. In what follows, meshes are designated by two integers separated by the multiplication sign "×" and followed by the type letter (A or B or C or D). Typically, the two integers indicate the numbers of divisions along the X- and Y-directions, respectively. Various boundary conditions are considered in the following sub-sections.

3.1.1. Free Boundary Conditions

For this case, no constraints are applied to the lateral displacement W and to the rotations about X- and Y-axes. Therefore, three zero natural frequencies exist. This is confirmed by the present results in which the first three natural frequencies are technically zero, and no spurious modes are detected. The first five non-zero natural frequencies are listed in Table **3.1**, where the dimensionless frequency parameter is $\Omega = \omega a^2 \sqrt{\frac{\rho_0 h}{D}}$ with a being the side length of the plate and $D = \frac{Eh^3}{12(1-v^2)}$ being the flexural rigidity. Solutions from [3.1] are also included for comparison.

3.1.2. Simply Supported at Four Edges

Boundary conditions in this case are, $W = \Theta_Y = 0$ for sides parallel to X-axis, and $W = \Theta_X = 0$ for sides parallel to Y-axis. The computed frequency parameters with analytical solutions from [3.2] are presented in Table **3.2**.

3.1.3. Clamped at Four Sides

In this case, the global DOF, W, Θ_X and Θ_Y of nodes on the four sides are constrained to zero. Computed results are listed in Table **3.3**.

3.1.4. Convergence and Discussion

The convergence issue is studied by making use of the relationship between the normalized (with respect to analytical or known solution) first elastic natural frequencies and effective DOF, or NEQ. Computed results are presented in Fig. **3.2**.

With reference to the foregoing tables, it is observed that the seventh and eighth natural frequencies, that is, the fourth and fifth elastic modes, of the unconstrained plate, the second and third natural frequencies of the simply-supported and clamped plates are identical. Therefore, it is of interest to examine the characteristics of these identical natural frequencies. To this end, their corresponding mode shapes are plotted in Fig. **3.3**.

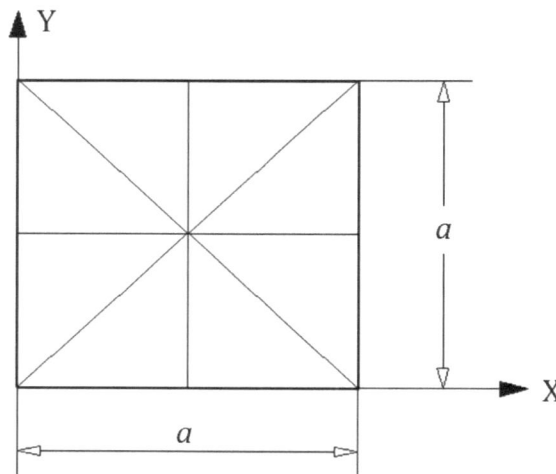

Figure 3.1: Mesh 2×2C of square plate.

Table 3.1: First five non-zero frequency parameters $\Omega = \omega a^2 \sqrt{\frac{\rho_0 h}{D}}$ (completely free square plate).

Mesh	NEQ*	Ω_4	Ω_5	Ω_6	$\Omega_7 = \Omega_8$
4×4C	75	13.7921	20.1198	25.3210	39.7964
8×8C	243	13.5928	19.7481	24.5672	36.1462
12×12C	507	13.5289	19.6663	24.4051	35.4056
16×16C	867	13.5014	19.6351	24.3447	35.1323
20×20C	1323	13.4870	19.6200	24.3159	35.0013
24×24C	1875	13.4784	19.6117	24.3001	34.9283
28×28C	2523	13.4728	19.6066	24.2905	34.8834
32×32C	3267	13.4688	19.6033	24.2843	34.8537
4×4D	123	13.6886	20.1175	25.3164	36.9157
8×8D	435	13.5445	19.7396	24.5525	35.4127
12×12D	939	13.5007	19.6591	24.3932	35.0676
16×16D	1635	13.4824	19.6295	24.3354	34.9353
20×20D	2523	13.4728	19.6156	24.3085	34.8710
24×24D	3603	13.4671	19.6081	24.2940	34.8390
28×28D	4875	13.4633	19.6037	24.2854	34.8126
32×32D	6339	13.4605	19.6009	24.2800	34.7977
[3.1][1]		13.201	-	-	-
[3.1][2]		13.4728	19.5961	24.2702	34.8010

* NEQ denotes the number of total unknown displacements, or the effective DOF.
[1] Table 4.67, p.109; lower-bound solutions.
[2] Table 4.65, pp.104-106 and Figure 4.53, p.107; not necessarily lower-bound solutions.

Table 3.2: First five frequency parameters $\Omega = \omega a^2 \sqrt{\frac{\rho_0 h}{D}}$ (simply supported square plate).

Mesh	NEQ	Ω_1	$\Omega_2 = \Omega_3$	Ω_4	Ω_5
4×4C	39	21.8982	61.8968	123.833	1173.57
8×8C	175	20.2466	52.0618	87.4666	108.283
12×12C	407	19.9578	50.4886	82.5315	102.579
16×16C	735	19.8582	49.9633	80.8831	100.764
20×20C	1159	19.8125	49.7250	80.1373	99.9548
24×24C	1679	19.7878	49.5972	79.7381	99.5241
28×28C	2295	19.7731	49.5210	79.5005	99.2683
32×32C	3007	19.7636	49.4722	79.3484	99.1045
4×4D	87	20.2512	54.1891	87.5640	123.653

Table 3.2: contd….

8×8D	367	19.8621	50.4197	80.9517	104.024
12×12D	839	19.7912	49.7964	79.7952	100.888
16×16D	1503	19.7666	49.5838	79.3968	99.8448
20×20D	2359	19.7552	49.4870	79.2141	99.3739
24×24D	3407	19.7491	49.4352	79.1159	99.1236
28×28D	4647	19.7455	49.4045	79.0573	98.9759
32×32D	6079	19.7431	49.3850	79.0198	98.8824
[3.2][1]		19.74	49.35	78.96	98.70

[1] Table **11-4**, p.258, exact solution is $\Omega = \pi^2 \left[i^2 + \left(\frac{ja}{b} \right)^2 \right]$ for a square plate where i and j are mode numbers in the X- and Y-directions, respectively.

Table 3.3: First five frequency parameters $\Omega = \omega a^2 \sqrt{\frac{\rho_0 h}{D}}$ (clamped square plate).

Mesh	NEQ	Ω_1	$\Omega_2 = \Omega_3$	Ω_4	Ω_5
4×4C	27	396.070	590.799	1592.19	2491.43
8×8C	147	41.2240	89.3206	147.526	167.459
12×12C	363	37.8509	79.0208	122.762	144.460
16×16C	675	36.8451	76.0906	115.276	138.084
20×20C	1083	36.4613	74.9157	112.176	135.366
24×24C	1587	36.2819	74.3480	110.671	133.987
28×28C	2187	36.1840	74.0343	109.849	133.204
32×32C	2883	36.1247	73.8433	109.355	132.721
4×4D	75	41.8442	101.672	149.081	1086.93
8×8D	339	37.3253	78.6610	117.237	149.002
12×12D	795	36.5517	75.5580	111.981	138.431
16×16D	1443	36.2855	74.5307	110.203	135.129
20×20D	2283	36.1638	74.0673	109.393	133.670
24×24D	3315	36.0985	73.8206	108.959	132.902
28×28D	4539	36.0598	73.6750	108.702	132.452
32×32D	5955	36.0352	73.5827	108.538	132.167
[3.1][1]		35.99	73.41	108.27	131.64
[3.1][2]		35.9852	73.3938	108.217	131.581

[1] Table **4.22**, p.60.
[2] Table **4.30**, p.66.

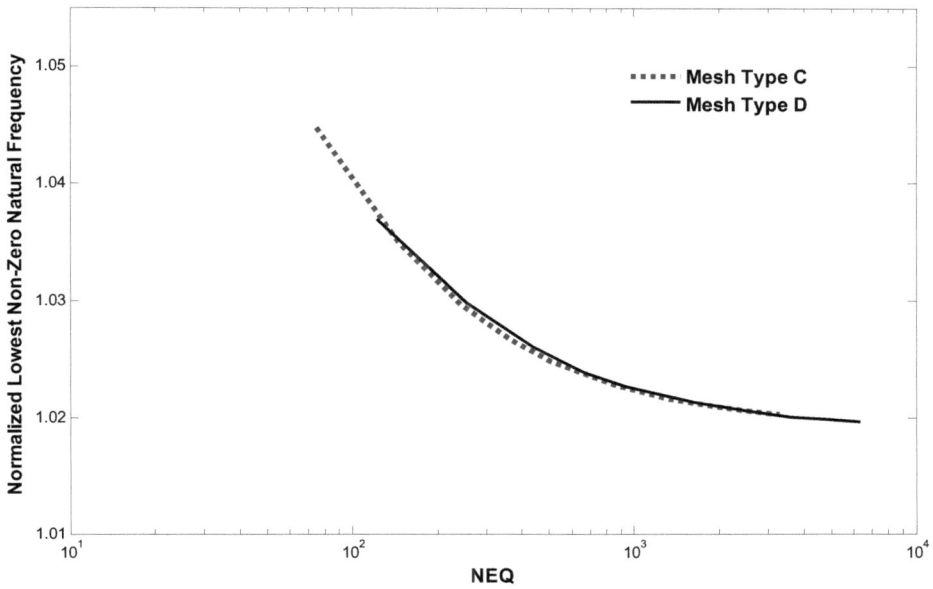

Figure 3.2(a): Convergence of computed fundamental frequency (completely free plate).

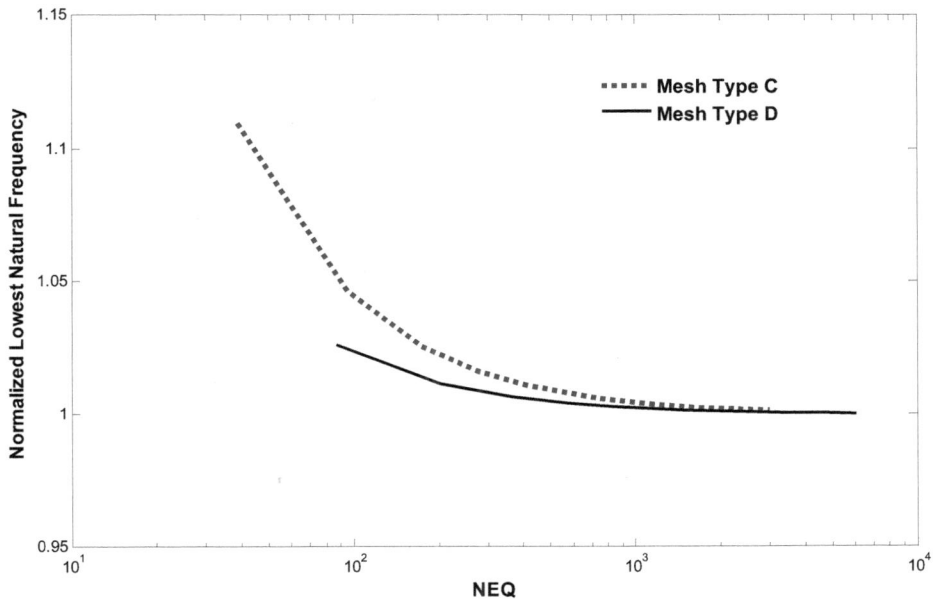

Figure 3.2(b): Convergence of computed fundamental frequency (simply supported plate).

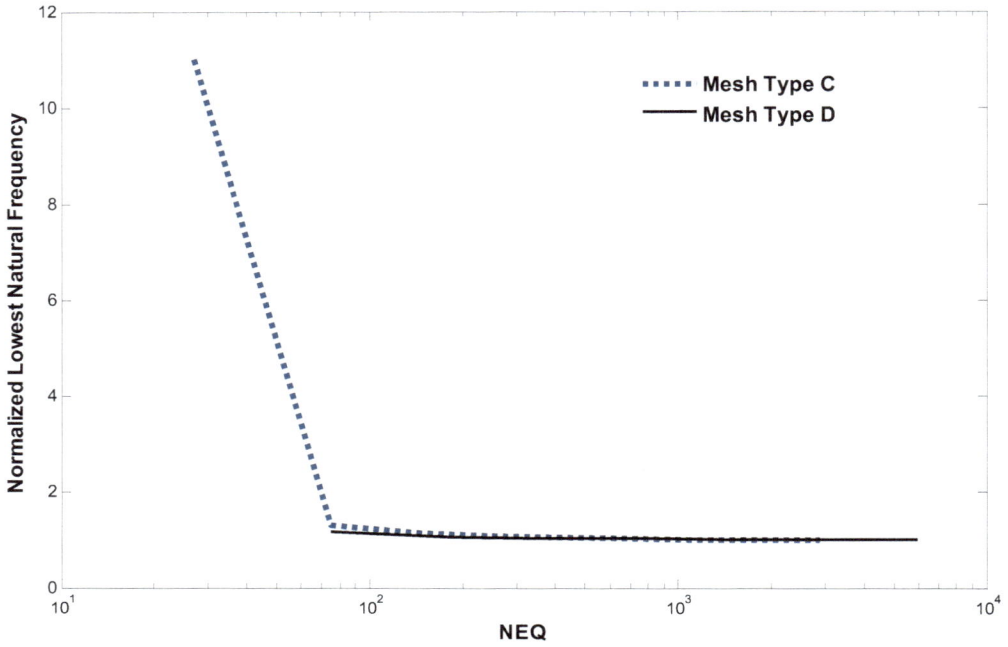

Figure 3.2(c): Convergence of computed fundamental frequency (clamped plate).

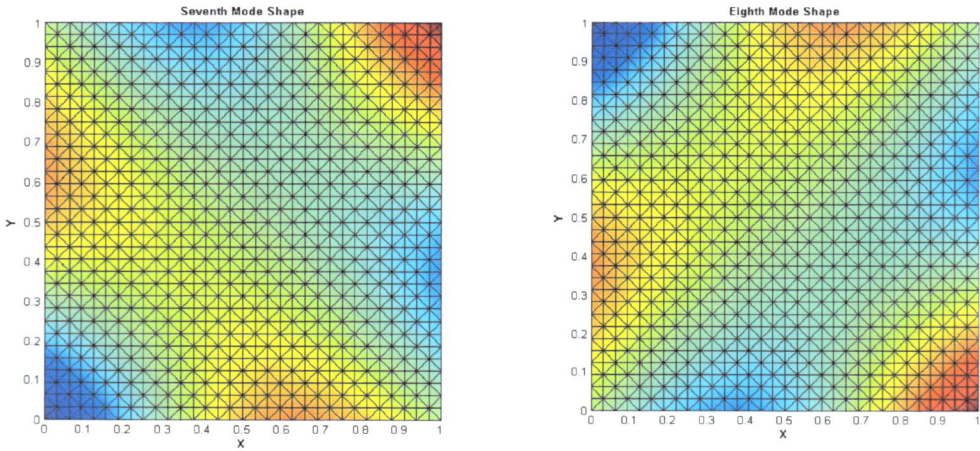

Figure 3.3(a): The seventh and eighth modes (fourth and fifth elastic modes) of completely free square plate.

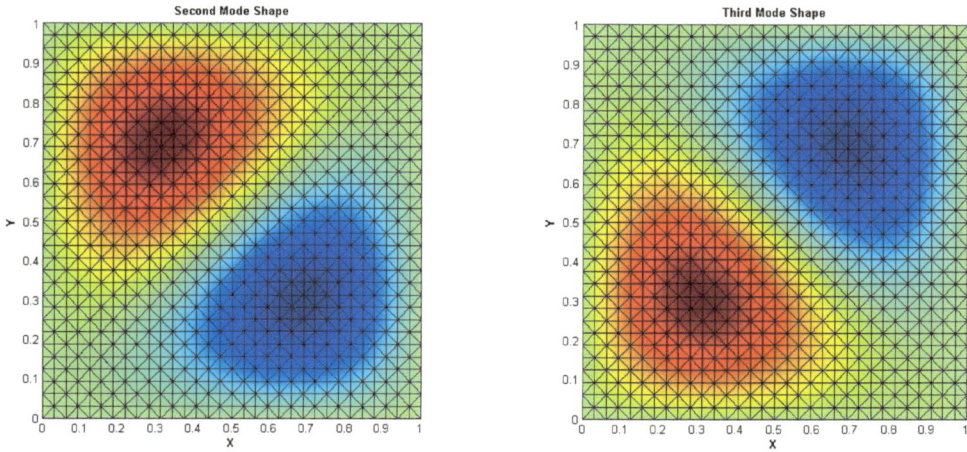

Figure 3.3(b): The second and third modes of simply supported square plate.

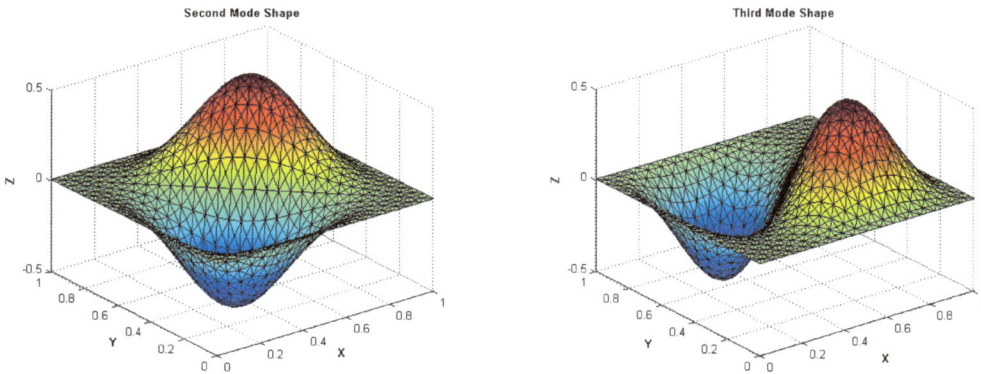

Figure 3.3(c): The second and third modes of clamped square plate.

Note that the mode shapes presented in Fig. **3.3** are produced by using the MATLAB® programming language [3.3], as are mode shapes to be presented in the remainder of this book. To produce a mode shape plot, the computed mode shape is added to the original configuration. This "new" configuration is plotted using the *patch* function of MATLAB® which plots polygons. To show more clearly the range and variation of displacements, color map is added to the plot of polygons. In MATLAB®, this is achieved by specifying the *FaceVertexCData* property when using the *patch* function. The *FaceVertexCData*, or face vertex color data, is typically set to be displacement W with the exception of the plane stress structure presented in Sec. 3.4 in which displacement V is the *FaceVertexCData*. As a result, the dark red color indicates the maximum displacement while the deep blue indicates the minimum displacement.

With reference to Fig. **3.3**, it is observed that the third mode shapes can be obtained by rotating the respective second mode shapes by 180° clockwise about the Z-axis. On the other hand, the eighth mode shape of the completely free plate is the result of rotating the seventh mode shape by 90 degrees, also clockwise.

Returning to the computed results for the convergence study presented in Fig. **3.2** some comments are in order.

(i)   It is evident that a consistent convergence pattern exists, which is approaching the correct solutions from above. That is, the present finite element provides upper bounds to the exact solutions.

(ii) Excellent accuracy (< 5% discrepancy) is achieved for the case of the completely free plate, even when the number of total unknowns, NEQ, is relatively low. Excellent accuracy is also achieved for the case of simply supported plate when NEQ is more than 100. This NEQ is observed to increase to more than 300 for the clamped plate.

(iii) Mesh type D outperforms type C only when NEQ is relatively low, typically up to a few hundreds, depending on boundary conditions. When NEQ is high, the two mesh types yield very small differences in computed results.

3.2. CIRCULAR PLATE AND EFFECT OF ASPECT RATIO

The objective in this section is to determine the range of radius-to-thickness ratio, or aspect ratio, over which the computed finite element results are accurate and reliable.

The circular plate is chosen because it has analytical solutions [3.4] of central deflections under uniform load, and available results [3.5] for free vibration analysis. It has been studied by other researchers for similar purpose [3.5-3.7].

According to [3.4], a circular plate of radius a and thickness h, with Young's modulus E and Poisson's ratio v, and under uniform transverse loading q, has the following central deflection,

$$\delta = \frac{qa^4}{64D}\left(1 + \frac{16}{5(1-v)}\frac{h^2}{a^2}\right)$$

if it is fully clamped, and

$$\delta = \frac{qa^4}{64D}\left(\frac{5+v}{1+v} + \frac{16}{5(1-v)}\frac{h^2}{a^2}\right)$$

if it is simply supported. In the above formulas $D = \frac{Eh^3}{12(1-v^2)}$ is the flexural rigidity.

In the finite element representation, only a quarter of the plate is discretized because of symmetry. Four meshes, 153A, 162F, 297D and 288F are applied. They are shown in Fig. **3.4**. These meshes are formed in such a way that 153A and 162F, and 297D and 288F, have comparable NEQ to each other. Since it is a plate bending problem, all the displacements along the X- and Y-directions and the DDOF are constrained. The sides of $X = 0$ and $Y = 0$ are sides of symmetry. Their boundary conditions are: $\Theta_Y = 0$ for the side $X = 0$; and $\Theta_X = 0$ for the side $Y = 0$. For the circular arc, the boundary conditions are: $W = \Theta_X = \Theta_Y = 0$ for the fully clamped case, and $W = 0$ for the simply supported case.

The radius of the plate is $a = 1.0$ m. Its material properties are: $E = 200$ GPa and $v = 0.25$. The intensity of uniform loading is $q = 1.0$ Pa. For various thicknesses the normalized central displacements are plotted in Fig. **3.5** for the case of simply supported plate, and Fig. **3.6** for the clamped plate.

It is seen from Figs. **3.5** and **3.6** that,

(i)   there is only small difference in the performance among the meshes, with the exception of 297D, in the moderately-thick to thick regime where the radius-to-thickness ratio is equal to and less than 10. In the thin regime, however, the four meshes perform from poorly (in the case of clamped plate) or fairly (in the case of simply supported plate) to very well;

(ii) mesh 297D seems to be very well suited for the problem. The calculated normalized central displacements fall within ±2% of unity, over a wide range of radius-to-thickness ratio, 10 to 10,000, and for both boundary conditions. If a 10% discrepancy were to be allowed, the radius-to-thickness ratio could be extended to 4 for the clamped plate and 2 for the simply supported plate;

(iii) the computed displacements are lower bound solutions in the thin regime, but become upper bound solutions in the moderately-thick to thick regime. The transition, in terms of the value of the aspect ratio, seems to occur at 10 for both cases of boundary conditions; and

(iv) the excellent performance of the 297D mesh is attributed to, sufficient elements around the center of plate and type D mesh along the boundary. Mesh 288F has very similar NEQ, but performs as well as 297D only over a small range of radius-to-thickness ratio: 10 to 200.

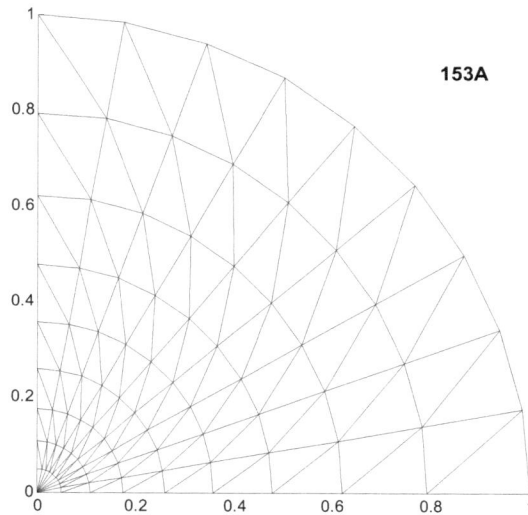

Figure 3.4(a): Mesh 153A used in circular plate.

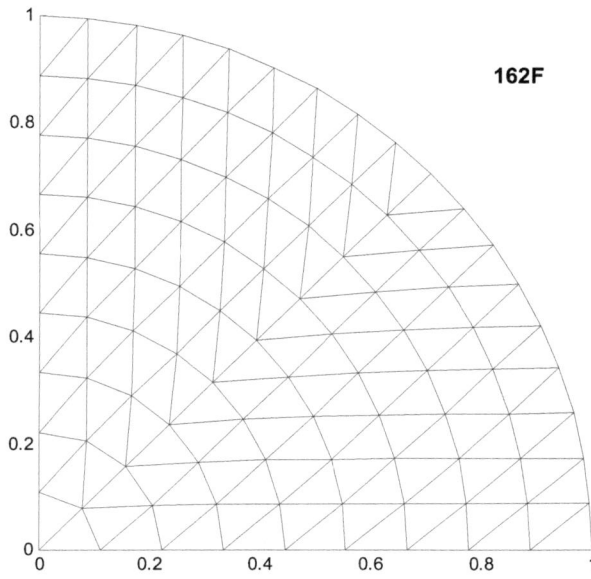

Figure 3.4(b): Mesh 162F used in circular plate.

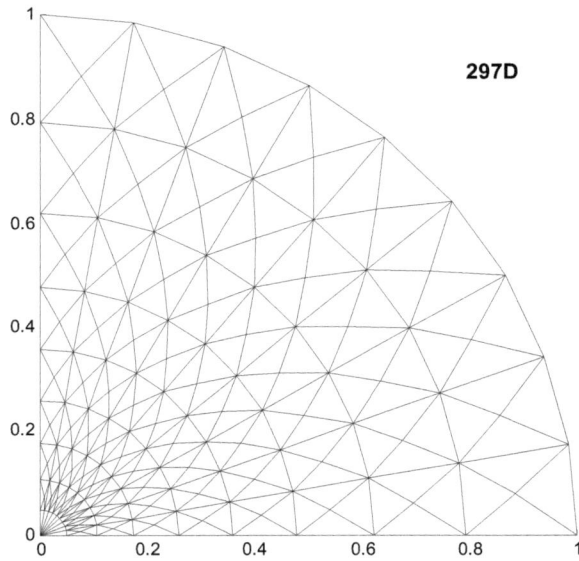

Figure 3.4(c): Mesh 297D used in circular plate.

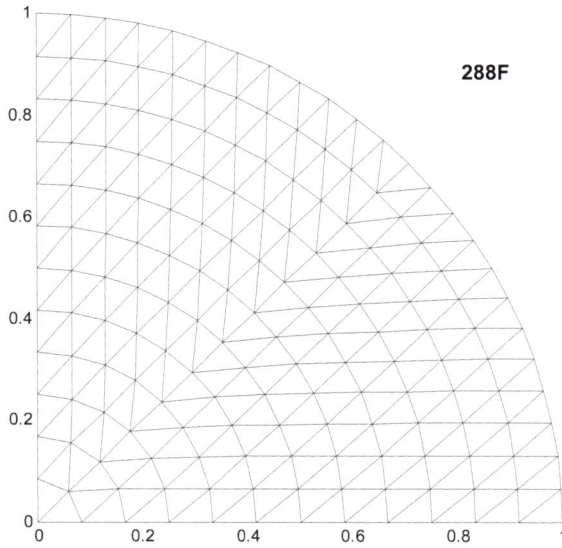

Figure 3.4(d): Mesh 288F used in circular plate.

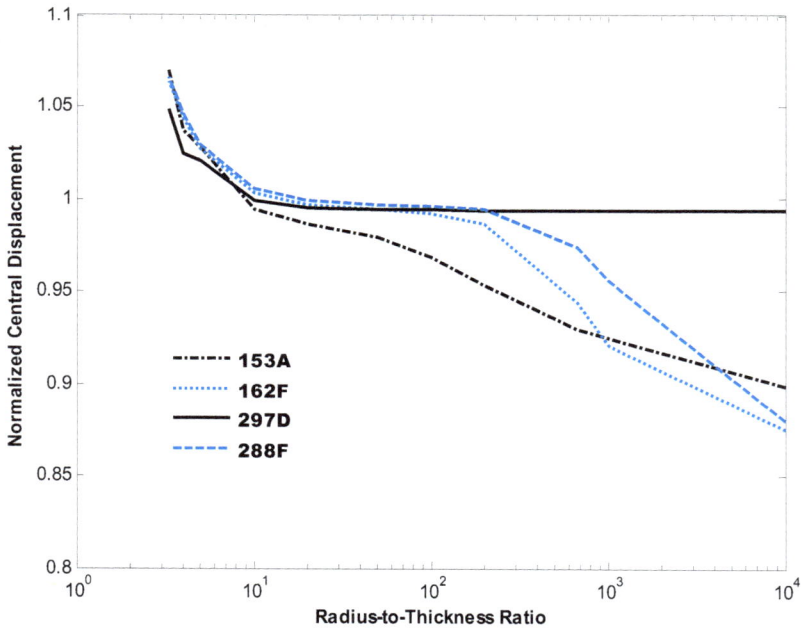

Figure 3.5: Displacement *versus* radius-to-thickness ratio for simply supported plate.

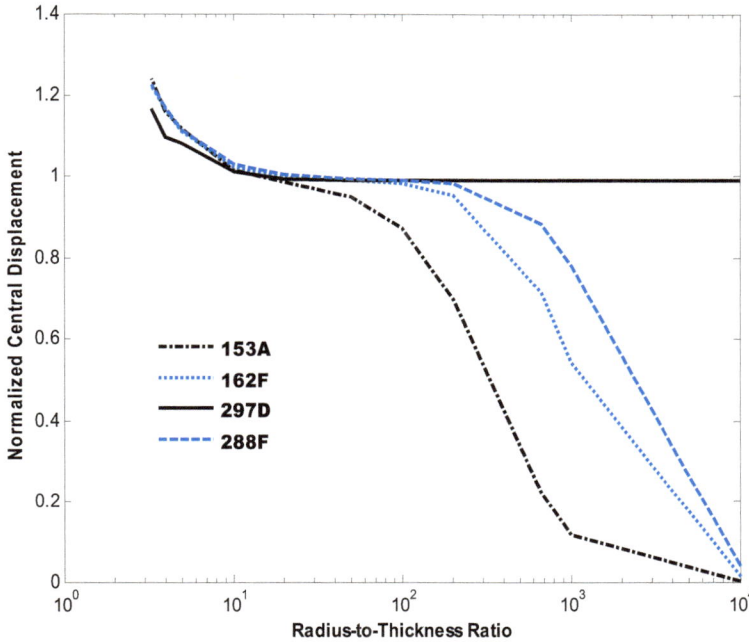

Figure 3.6: Displacement *versus* radius-to-thickness ratio for clamped plate.

For free vibration analysis, the 153A and 297D meshes are used, since they represent relatively poor and very good discretizations. Symmetry boundary conditions are applied along the sides $X = 0$ and $Y = 0$, due to the fact that only the fundamental frequency is of concern here which is associated with an axisymmetric mode of vibration. The circular arc is either clamped or simply supported. That is, the same boundary conditions as in the static analysis are applied. The radius of the plate remains at $a = 1.0$ m. Material properties are: $E = 200$ GPa, $v = 0.3$ and $\rho_0 = 7830$ kg/m^3. The Poisson's ratio is given a different value for direct comparison with results of [3.5], which found that, the fundamental frequency parameter $\Omega_1 = \omega_1 a^2 \sqrt{\frac{\rho_0 h}{D}}$ has the following values:

For a simply supported circular plate, $\Omega_1 = 4.9360$ for $h/a \leq 0.01$, and $\Omega_1 = 4.8975, 4.7876$ and 4.6234 for $h/a = 0.1, 0.2$ and 0.3, respectively; and

For a clamped circular plate, $\Omega_1 = 10.250$ for $h/a \leq 0.01$, and $\Omega_1 = 9.9909, 9.3225$ and 8.4676 for $h/a = 0.1, 0.2$ and 0.3, respectively.

The normalized computed fundamental frequencies are plotted against the radius-to-thickness ratios in Figs. **3.7** and **3.8**. The following observations are in order.

(i)   In contrast to the static analysis, the computed frequencies are upper bound solutions in the thin regime, but lower bound solutions in the moderately-thick to thick regime. The transition point seems to be about 30 for the simply supported case and 10 for the clamped case.

(ii) Mesh 297D performs very well, over the entire range of radius-to-thickness ratios considered, and for both boundary conditions. It may be worth recalling that this mesh was not accurate in predicting the central displacement of a clamped thick plate, see Fig. **3.7**.

(iii) When the boundary condition is simple support, mesh 153A yields surprisingly satisfactory results over the entire range of aspect ratios considered. For the clamped boundary condition, the range is reduced to 3 and 20. In either case, the satisfactory range of aspect ratio is wider than the statics counterpart. This seems to suggest that a mesh that is satisfactory for static analysis is also satisfactory for free vibration analysis.

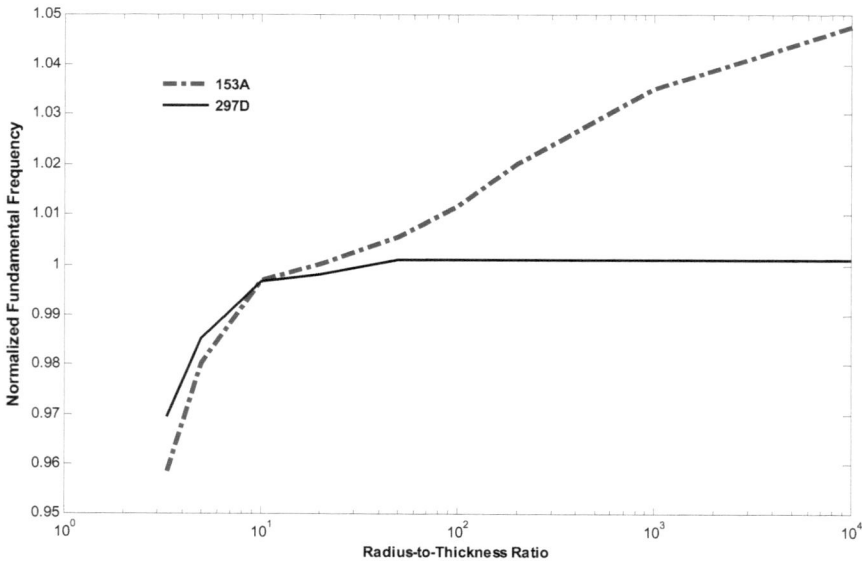

Figure 3.7: Fundamental frequency *versus* radius-to-thickness ratio for simply supported plate.

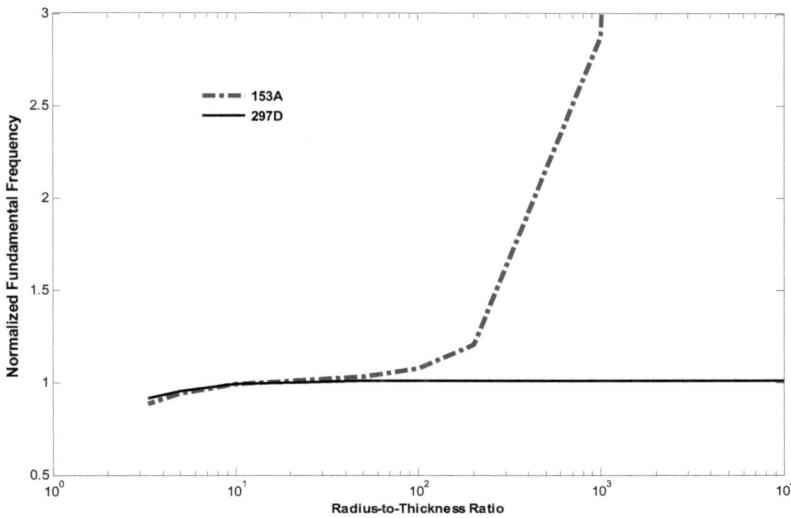

Figure 3.8: Fundamental frequency *versus* radius-to-thickness ratio for clamped plate.

3.3. SKEW PLATES

The skew plate shown in Fig. **3.9** has a skew angle of $30°$. The side length is a = 1.0 m, and thickness is h = 0.005 m. The material properties are: E = 200 GPa, v = 0.3 and ρ_0 = 7830 kg/m^3. Similar to the square plate in the previous section, the in-plane displacements and rotation about the Z-axis are constrained. At the clamped side where $Y = 0$, constraints $W = \Theta_X = \Theta_Y = 0$ are also imposed. The finite element results are presented in Table **3.4**, which are computed from various type B meshes.

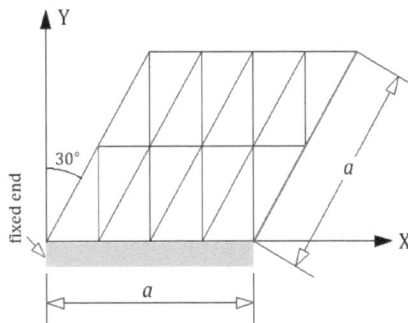

Figure 3.9: Representative mesh of skew plate.

Table 3.4: First five frequency parameters $\Omega = \omega a^2 \sqrt{\frac{\rho_0 h}{D}}$ of skew plate.

Mesh	NEQ	Ω_1	Ω_2	Ω_3	Ω_4	Ω_5
4×3B	45	4.58632	25.7584	38.8591	83.4569	165.736
6×4B	84	4.47035	18.5942	33.7038	60.0117	105.281
8×5B	135	4.34443	14.7553	31.2490	48.5701	72.2013
10×6B	198	4.23451	12.7520	29.7680	41.3058	58.1313
12×8B	312	4.11241	11.2502	28.2297	34.7561	49.4233
16×10B	510	4.03097	10.3801	27.1605	30.3027	45.5565
20×12B	756	3.98940	9.97011	26.5158	28.3544	43.8155
24×15B	1125	3.96270	9.71869	26.0394	27.2286	42.7277
28×18B	1566	3.94955	9.59507	25.7791	26.6874	42.1917
32×20B	1980	3.94361	9.53870	25.6550	26.4461	41.9506
[3.1][1]		3.961	10.19	-	-	-
[3.1][2]		3.95	9.65	25.5	26.1	42.4

[1] Table **5.12**, p.170.
[2] Table **5.14**, p.172.

Two sets of results are referenced from [3.1]. Both sets were obtained by using the Rayleigh-Ritz method with 18 or more terms of the deflection function which are products of characteristic beam functions. They are therefore upper bound solutions. Considering Ω_1, the two finest meshes employed, 28×18B and 32×20B, yield results that are slightly lower than the values of 3.961 or 3.95, both from [3.1].

Compared with results from 32×20B which is the finest mesh used, it is found that mesh 20×12B (NEQ = 756) gives less than 5% of discrepancies on the first five frequencies; Mesh 24×15B, having a NEQ of 1125, yields less than 5% of discrepancies, on all ten frequencies. The first five mode shapes of vibration are given in Fig. **3.10**.

3.4. PLANE STRESS OR MEMBRANE STRUCTURE

While the structure considered in this section is strictly not a plate it is included here for two main reasons. First, it is one of the main components of the shell element. Second, for completeness its natural frequencies and mode shapes are of interest to the engineers in general.

First Mode Shape

Fixed End

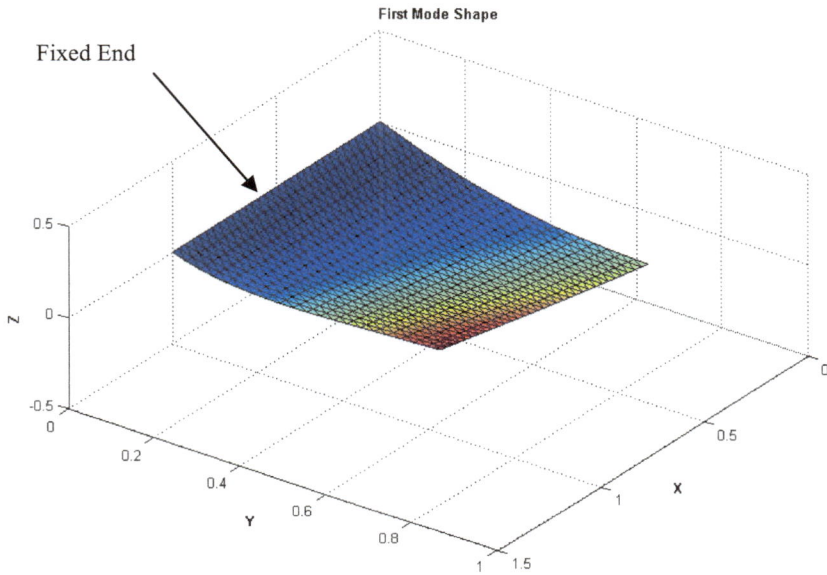

Figure 3.10(a): First mode shape of skew plate.

Second Mode Shape

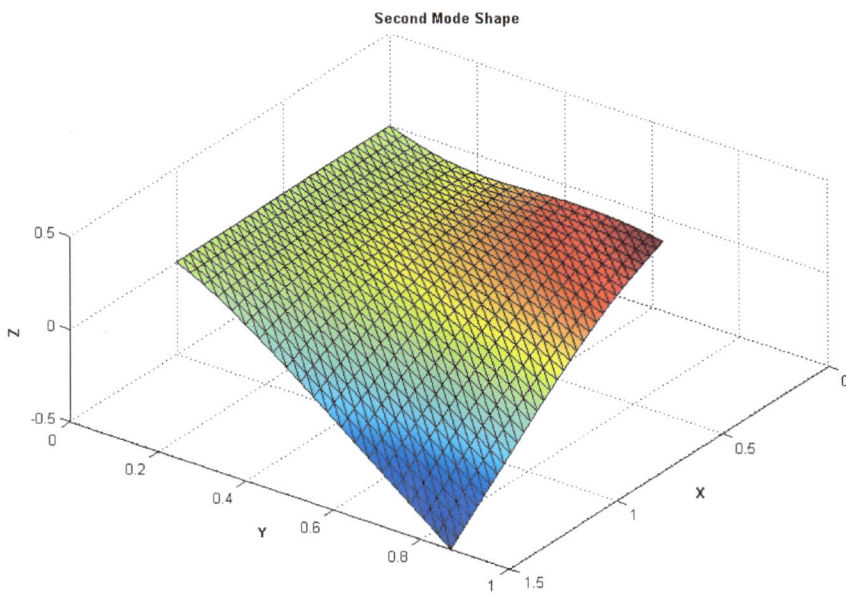

Figure 3.10(b): Second mode shape of skew plate.

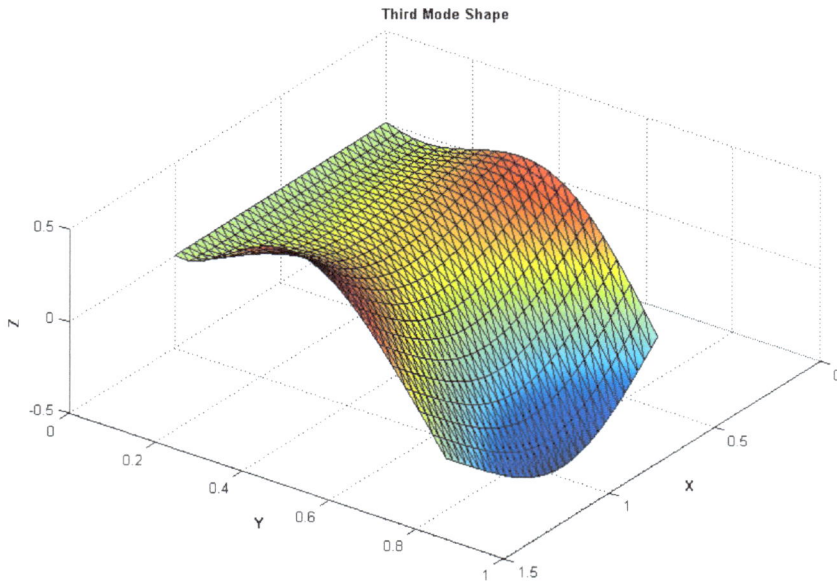

Figure 3.10(c): Third mode shape of skew plate.

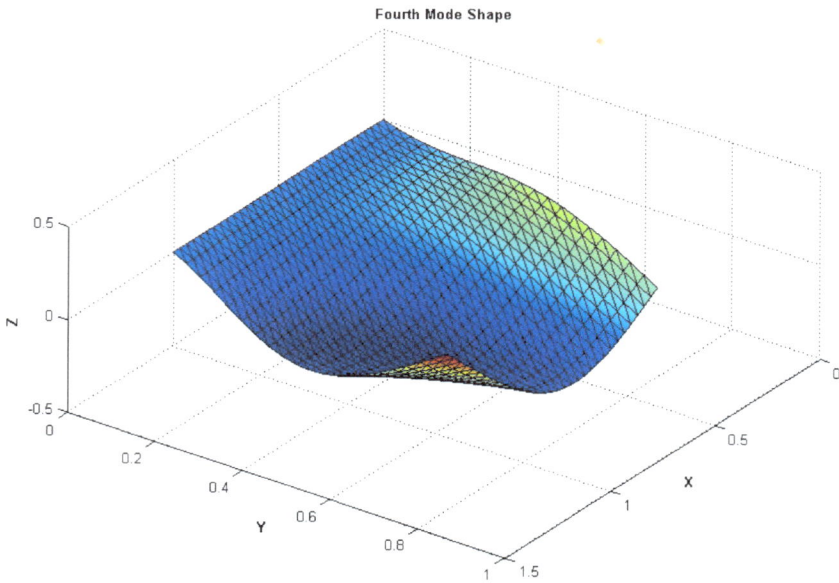

Figure 3.10(d): Fourth mode shape of skew plate.

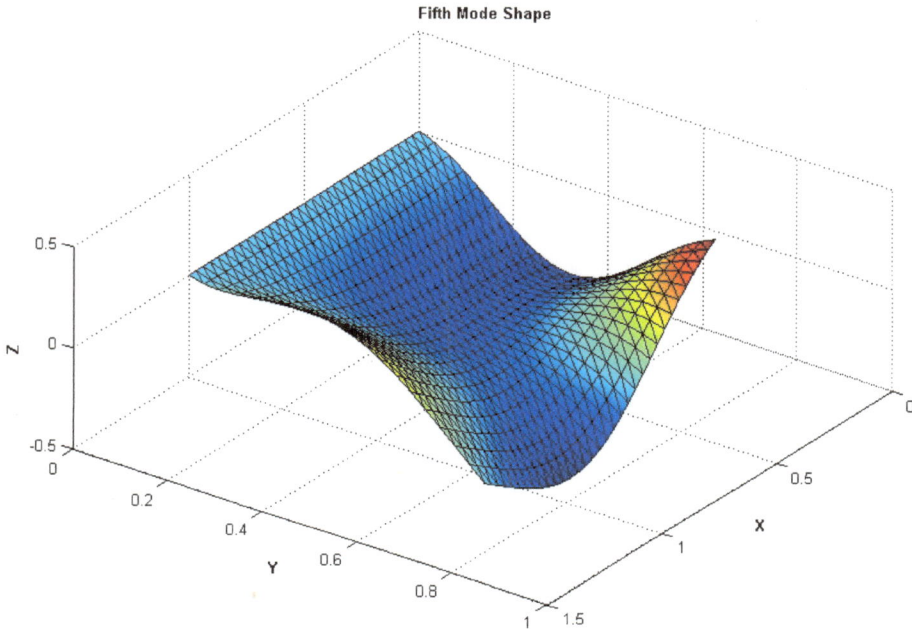

Figure 3.10(e): Fifth mode shape of Skew plate.

For the plane stress or membrane structure shown in Fig. **3.11**, its deformation takes place on the *X-Y* plane only. Therefore, W, Θ_X and Θ_Y are constrained at all nodes. At the fixed side where $X = 0$, constraints $U = V = \Theta_Z = 0$ are also imposed. In addition to the length of $L = 1.2192$ m (48.0 in) and height of $H = 0.3048$ m (12.0 in) given in Fig. **3.11**, the material properties of the structure are: $E = 207$ GPa (30×10^6 psi), $v = 0.25$ and $\rho_0 = 7795$ kg/m^3 (0.000730 lb·sec^2/in^4), and the thickness (or the dimension in the *Z*-direction) is $h = 25.4$ mm (1.0 in). The computed finite element results are presented in Table **3.5**. The first ten mode shapes associated with the 32×8D mesh are included in Fig. **3.12**. In the mode shape plots, the dark red indicates maximum vertical displacement *V* and the deep blue indicates the minimum vertical displacement *V*.

The following observations are in order.

(i)   The convergence pattern of approaching from above (see Sec. 3.1) is observed for all ten natural frequencies.

(ii) The coarsest mesh used, 8×2D, whose NEQ is only 120, is in fact sufficiently accurate. Compared with frequencies computed by the 32×8D mesh, the discrepancies are, from the first to the tenth modes, 0.8977%, 1.010%, 0.2237%, 1.104%, 0.9165%, 0.8456%, 0.6093%, 2.083%, 0.2037% and 0.3194%, respectively. The largest discrepancy is associated with the eighth mode.

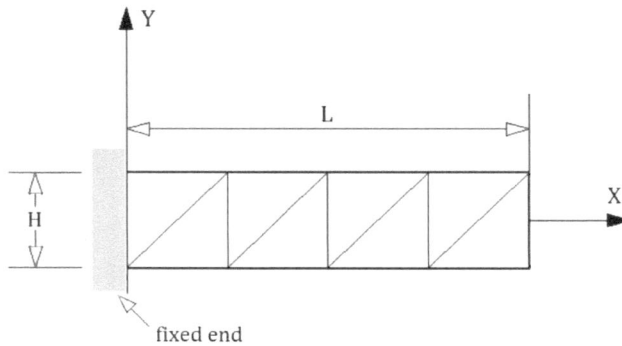

Figure 3.11: Representative mesh of plane stress or membrane structure.

Table 3.5: First ten natural frequencies (in Hz) of plane stress structure.

Mesh	NEQ	f_1	f_2	f_3	f_4	f_5
8×2D	120	165.032	843.703	1060.60	1945.82	3143.77
16×4D	432	163.952	837.440	1058.88	1930.76	3125.00
20×5D	660	163.781	836.391	1058.60	1928.13	3120.96
24×6D	936	163.678	835.743	1058.43	1926.48	3118.34
28×7D	1260	163.611	835.311	1058.32	1925.37	3116.53
32×8D	1632	163.563	835.005	1058.24	1924.58	3115.22
Mesh	NEQ	f_6	f_7	f_8	f_9	f_{10}
8×2D	120	3192.18	4376.01	5333.78	5391.36	6002.58
16×4D	432	3171.34	4362.40	5247.61	5393.14	5999.19
20×5D	660	3168.59	4357.42	5236.88	5388.56	5993.42
24×6D	936	3167.03	4353.90	5230.96	5385.00	5989.03
28×7D	1260	3166.07	4351.37	5227.33	5382.37	5985.83
32×8D	1632	3165.42	4349.51	5224.94	5380.40	5983.48

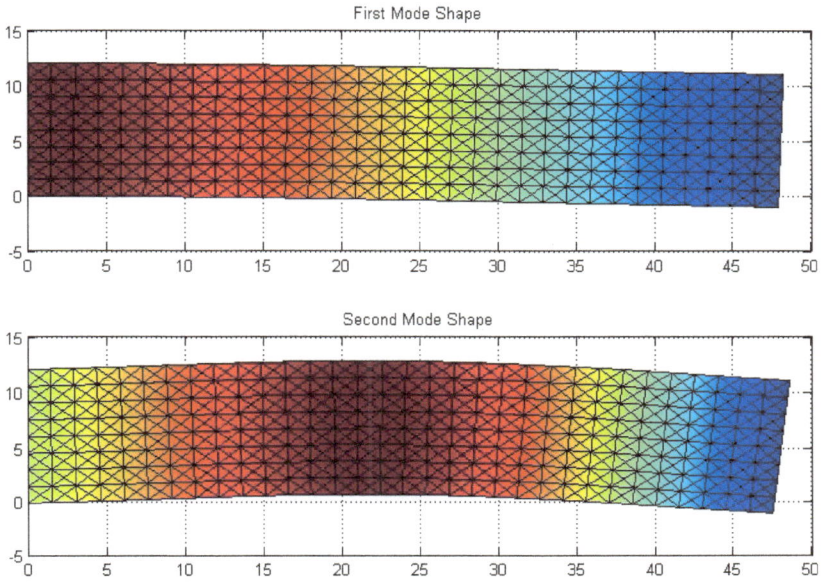

Figure 3.12(a): First and second mode shapes of plane stress structure.

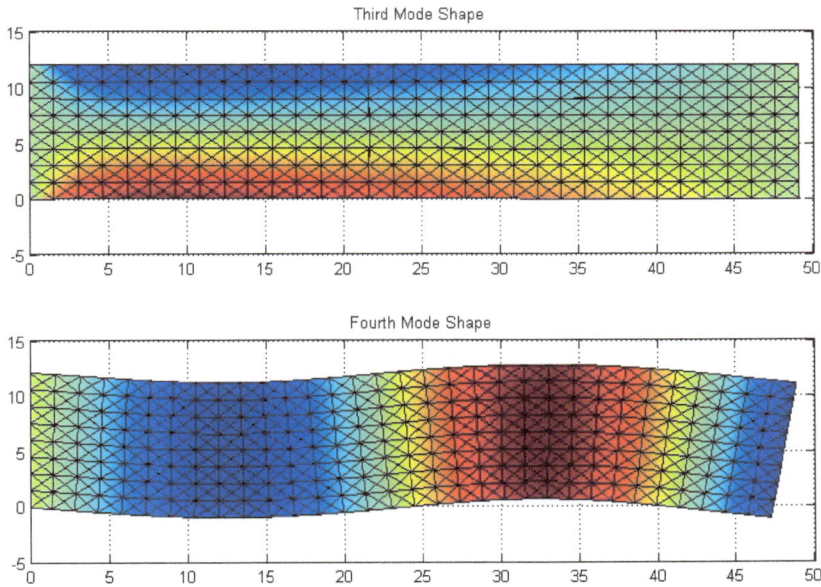

Figure 3.12(b): Third and fourth mode shapes of plane stress structure.

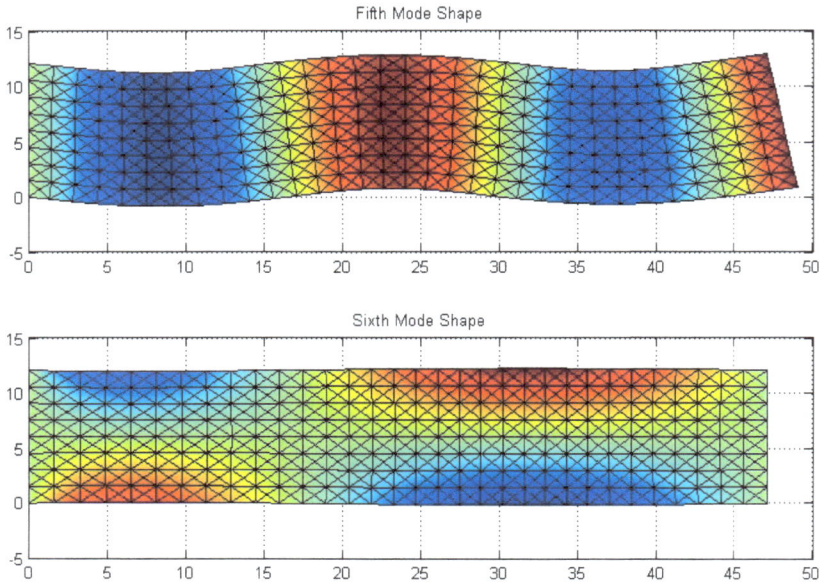

Figure 3.12(c): Fifth and sixth mode shapes of plane stress structure.

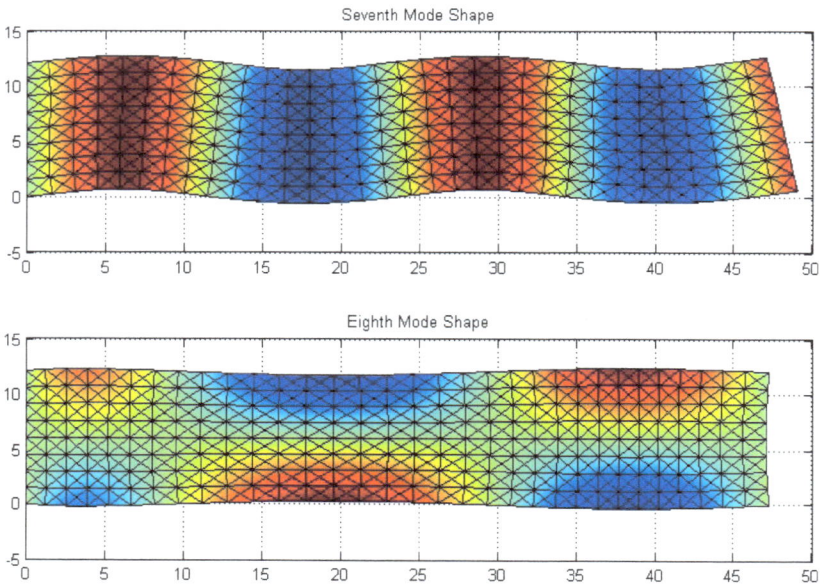

Figure 3.12(d): Seventh and eighth mode shapes of plane stress structure.

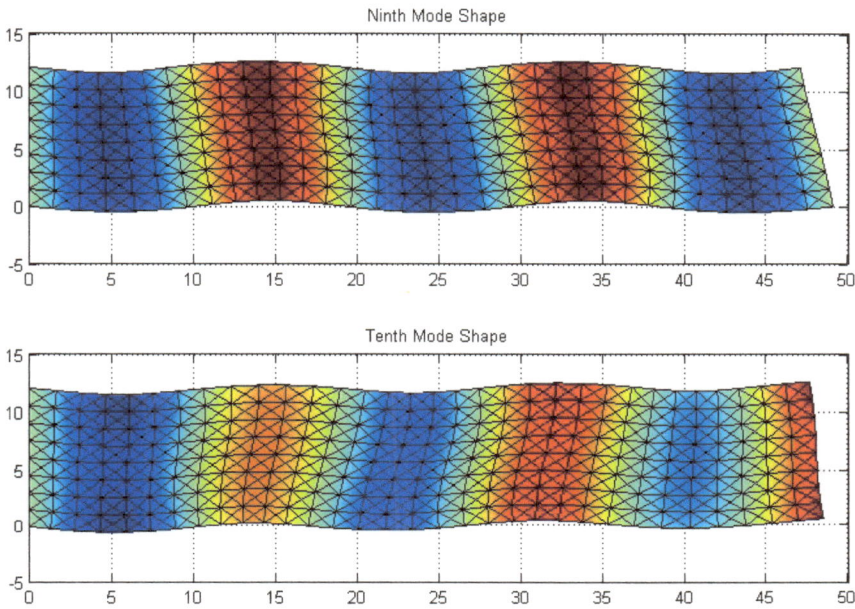

Figure 3.12(e): Ninth and tenth mode shapes of plane stress structure.

REFERENCES

[3.1]   A.W. Leissa, *Vibration of Plates*, reprint. New York: Acoustical Society of America, 1993.

[3.2]   R.D. Blevins, *Formulas for Natural Frequency and Mode Shape*, New York: Van Nostrand Reinhold, 1979.

[3.3]   MathWorks, *MATLAB, The Language of Technical Computing*, Natick, Massachusetts, U.S.A., 2004.

[3.4]   S. Timoshenko, and S. Woinowsky-Krieger, *Theory of Plates and Shells*, 2nd ed. New York: McGraw-Hill, 1956.

[3.5]   K.M. Liew, and B. Yang, "Three-dimensional vibrations elasticity solutions for free vibrations of circular plates: a polynomials-Ritz analysis", *Comp. Methods Appl. Mech. Engrg.*, vol. 175, pp. 189-201, June 1999.

[3.6]   N. Carpenter, T. Belystchko, and H. Stolarski, "Locking and shear scaling factors in C^0 bending elements", *Comp. Struct.*, vol. 22, pp. 39-52, February 1986.

[3.7]   O.C. Zienkiewicz, R.L. Taylor, P. Papadopoulos, and E. Onate, "Plate bending elements with discrete constraints: New triangular elements", *Comp. Struct.*, vol. 35, pp. 505-522, August 1990.

Send Orders for Reprints to reprints@benthamscience.net
Vibration and Nonlinear Dynamics of Plates and Shells, 2014, 55-76   **55**

CHAPTER 4

Vibration Analysis of Shells with Single Curvature

Abstract: This chapter is concerned with vibration analysis of shell structures having single curvature. The latter include cylindrical curved panel with rectangular projection, cylindrical curved panel with trapezoidal projection, the Scordelis-Lo roof, and cylindrical shell clamped at both ends. In this latter case the effect of aspect ratio is also studied.

Keywords: Vibration analysis, cylindrical, panels, Scordelis-Lo roof, shells.

4.1. CYLINDRICALLY CURVED PANEL WITH RECTANGULAR PROJECTION

As illustrated in Fig. **4.1**, the cylindrical panel has a radius of R = 609.6 mm (24 in), a circumferential width of L_c = 304.8 mm (12 in), and a longitudinal length of L = 304.8 mm (12 in). The thickness of the panel is h = 3.048 mm (0.12 in). In finite element modeling, the entire panel is discretized. For nodes located at the clamped base, all nodal DOF are constrained to zero. The material properties are: E = 207 GPa (30×10^6 psi), v = 0.3 and ρ_0 = 7850 kg/m^3 (0.000735 lb·sec^2/in^4). Computed frequencies are listed in Table **4.1** where the frequency parameter is defined by $\Omega = \omega R \sqrt{\rho_0(1 - v^2)/E}$. For mesh designation, the first integer indicates the number of divisions in the circumferential direction, and the second the number of divisions in the longitudinal direction. In Table **4.1** the finite element results from [4.1] are included for comparison. For the ten frequencies presented, very close agreements among those of [4.1] and those with meshes 12×12D and finer can be observed. Note that these meshes have NEQ of 1800 and higher. On the other hand, compared with results from the 28×28D mesh, the 8×8D mesh having NEQ = 816, yields discrepancy less than 5%. Using mesh 12×12D (NEQ = 1800) the discrepancy is less than 2%. Finally, examination of the mode shapes indicates that they qualitatively resemble those of a cylindrical panel with trapezoidal projection, which are presented in Fig. **4.3** in the next section.

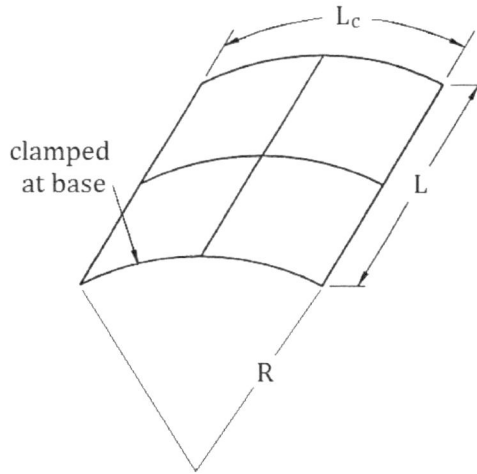

Figure 4.1: Clamped cylindrical panel with rectangular projection.

Table 4.1: First ten frequency parameters $\Omega = \omega R \sqrt{\frac{\rho_0 (1-\nu^2)}{E}}$ of cylindrical panel.

Mesh	NEQ	Ω_1	Ω_2	Ω_3	Ω_4	Ω_5
4×4D	216	0.0628525	0.101795	0.186689	0.265323	0.307530
8×8D	816	0.0617779	0.0996035	0.178177	0.250174	0.284293
12×12D	1800	0.0614196	0.0990335	0.176737	0.246435	0.278802
16×16D	3168	0.0612719	0.0988093	0.176237	0.245014	0.276746
20×20D	4920	0.0611971	0.0986986	0.176009	0.244342	0.275787
24×24D	7056	0.0611534	0.0986356	0.175887	0.243976	0.275271
28×28D	9576	0.0611249	0.0985958	0.175813	0.243754	0.274962
[4.1]		0.0611553	0.0985506	0.175869	0.244401	0.275131
Mesh	NEQ	Ω_6	Ω_7	Ω_8	Ω_9	Ω_{10}
4×4D	216	0.436501	0.576263	0.635639	0.657979	0.682809
8×8D	816	0.389000	0.538586	0.546726	0.570935	0.595773
12×12D	1800	0.381286	0.527787	0.529253	0.557352	0.580590
16×16D	3168	0.378633	0.522891	0.523117	0.553149	0.575188
20×20D	4920	0.377455	0.519999	0.520856	0.551335	0.572717
24×24D	7056	0.376848	0.518491	0.519645	0.550390	0.571408
28×28D	9576	0.376501	0.517620	0.518936	0.549832	0.570634
[4.1]		0.376694	0.517175	0.518614	0.549537	0.571460

4.2. CYLINDRICALLY CURVED PANEL WITH TRAPEZOIDAL PROJECTION

The cylindrically curved panel with trapezoidal projection considered in this section is the fan blade shown in Fig. **4.2**. The circumferential width tapers linearly from $L_b = 304.8$ mm (12 in) at the base to $L_e = 152.4$ mm (6 in) at the free end. The blade has a uniform thickness $h = 3.175$ mm (0.125 in), a length of $L = 304.8$ mm (12 in), and a constant radius $R = 304.8$ mm (12 in). Material properties are: $E = 207$ GPa (30×10^6 psi), $\nu = 0.3$ and $\rho_0 = 7850$ kg/m^3 (0.000735 lb·sec^2/in^4). For finite element computation, the entire panel is discretized. At the clamped base, all nodal DOF are set to zero. The computed frequency parameters $\Omega = \omega R \sqrt{\rho_0 (1 - \nu^2) / E}$ are listed in Table **4.2**. They are compared with the finite element results from Refs. [4.1, 4.2]. With the exception of Ω_1 in [4.2], both references gave higher values of frequencies than the presently obtained results. Since the present formulation yields upper bound solutions, it can be concluded that the presently obtained results are more accurate than those in [4.1, 4.2]. It is observed that for the first ten frequencies presented, mesh 16×16A or finer is required to provide results having a discrepancy of less than 5%. It is also observed that if accurate results of less than 5% are required only for the first five frequencies, mesh 12×12A with NEQ = 936 is sufficient. Overall, finer meshes than those of the cylindrical panel included in the last section are required. For completeness and identification purpose, the first ten mode shapes of the fan blade represented by the 28×28A mesh are presented in Fig. **4.3**. The mode shapes presented in Fig. **4.3** and the remainder of this book are produced by the MATLAB programming language, as are mode shapes that were presented in Chapter 3. To produce a mode shape plot, the computed mode shape is added to the original configuration. This "new" configuration is plotted using the *patch* function of MATLAB. To show more clearly the range and variation of displacements, color map is added based on displacement W. As a result, the dark red color indicates the maximum displacement while the deep blue indicates the minimum displacement. It may be appropriate to mention that the mode shapes are not plotted in scale.

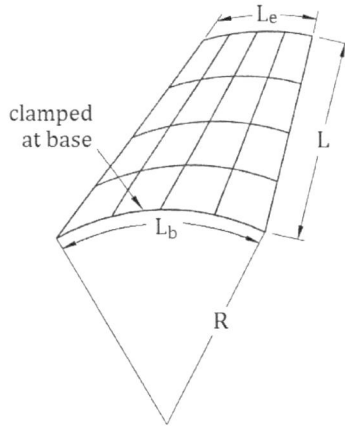

Figure 4.2: Fan blade as cantilevered cylindrical panel.

Table 4.2: First ten frequency parameters $\Omega = \omega R \sqrt{\frac{\rho_0 (1-\nu^2)}{E}}$ of fan blade.

Mesh	NEQ	Ω_1	Ω_2	Ω_3	Ω_4	Ω_5
8×8A	432	0.0720992	0.106327	0.181860	0.205448	0.309831
12×12A	936	0.0701526	0.104808	0.171328	0.194043	0.285290
16×16A	1632	0.0695788	0.104251	0.168682	0.191157	0.278080
20×20A	2520	0.0693278	0.103998	0.167734	0.190014	0.275435
24×24A	3600	0.0691910	0.103862	0.167295	0.189431	0.274214
28×28A	4872	0.0691055	0.103779	0.167052	0.189082	0.273544
32×32A	6336	0.0690468	0.103725	0.166901	0.188851	0.273128
[4.1]		0.0682542	0.109523	0.168595	0.194990	0.276879
[4.2]		0.0691727	0.104596	0.167716	0.190394	0.275302
Mesh	**NEQ**	Ω_6	Ω_7	Ω_8	Ω_9	Ω_{10}
8×8A	432	0.360218	0.397637	0.463981	0.520876	0.632058
12×12A	936	0.339141	0.347950	0.429631	0.472445	0.557323
16×16A	1632	0.331676	0.335388	0.420654	0.457913	0.529875
20×20A	2520	0.327719	0.331793	0.417291	0.452494	0.518871
24×24A	3600	0.325654	0.330322	0.415701	0.449979	0.513716
28×28A	4872	0.324495	0.329533	0.414821	0.448606	0.510932
32×32A	6336	0.323777	0.329046	0.414275	0.447761	0.509250
[4.1]		0.333777	0.340562	0.422801	0.461816	0.527688
[4.2]		0.328811	0.334115	0.420091	-	-

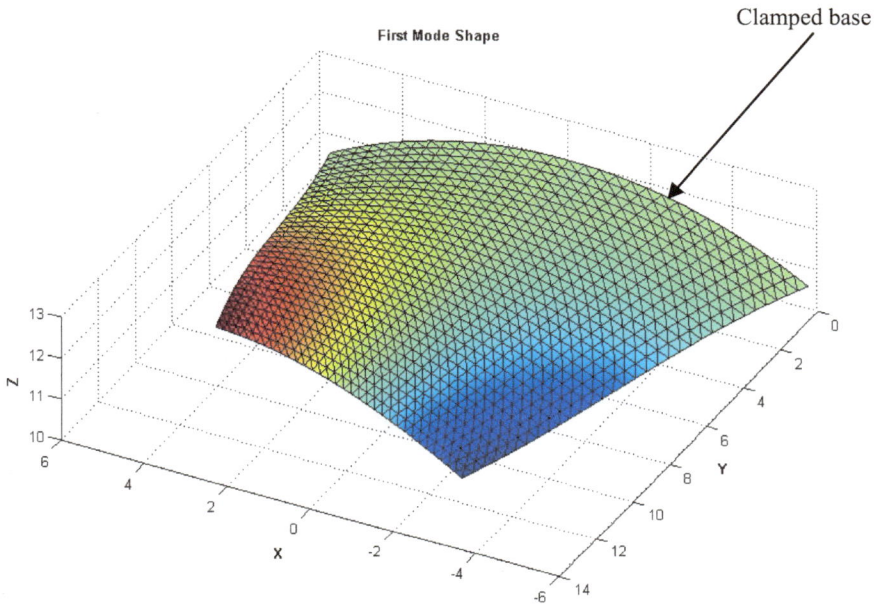

Figure 4.3(a): First mode shape of fan blade (Anti-symmetric mode).

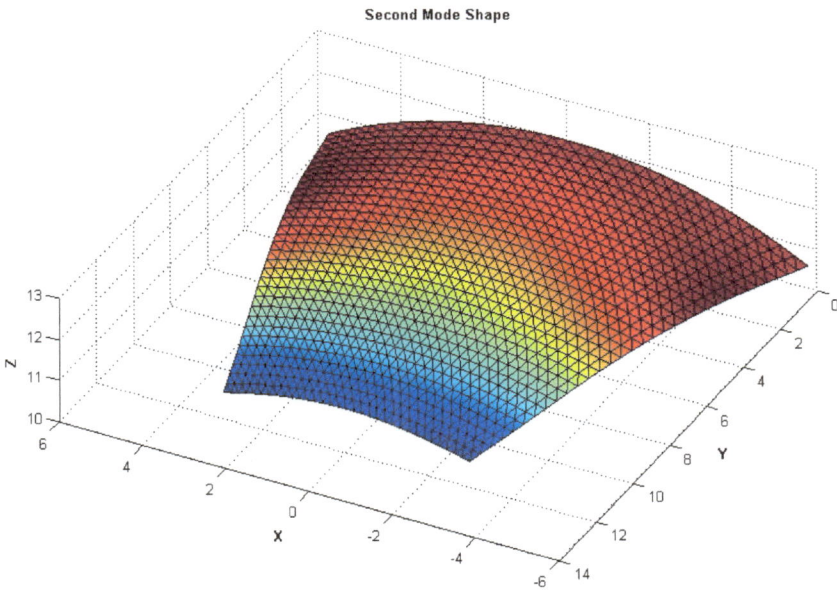

Figure 4.3(b): Second mode shape of fan blade (Symmetric mode).

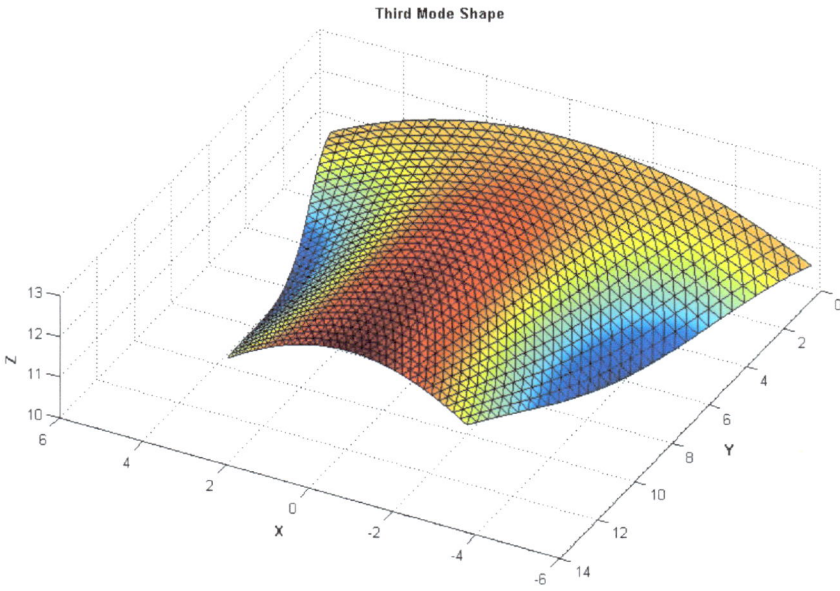

Figure 4.3(c): Third mode shape of fan blade (Symmetric mode).

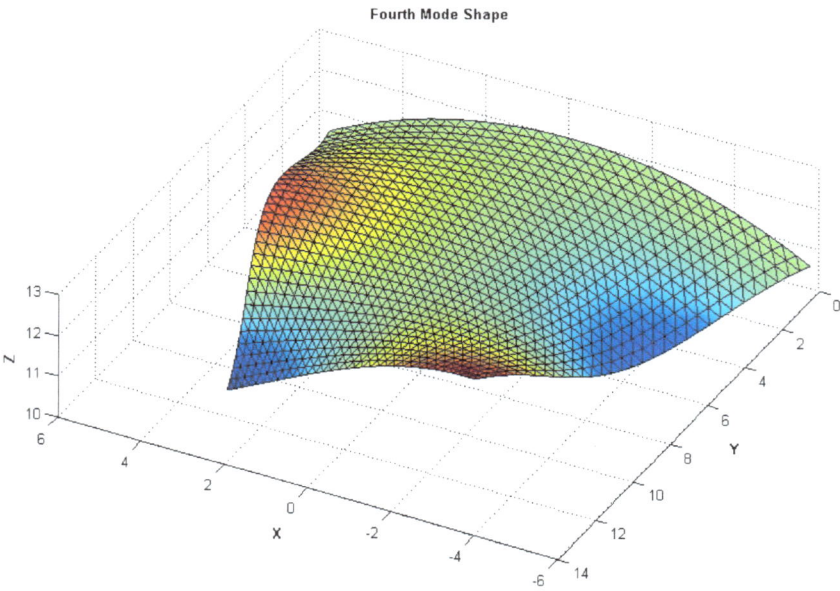

Figure 4.3(d): Fourth mode shape of fan blade (Anti-symmetric mode).

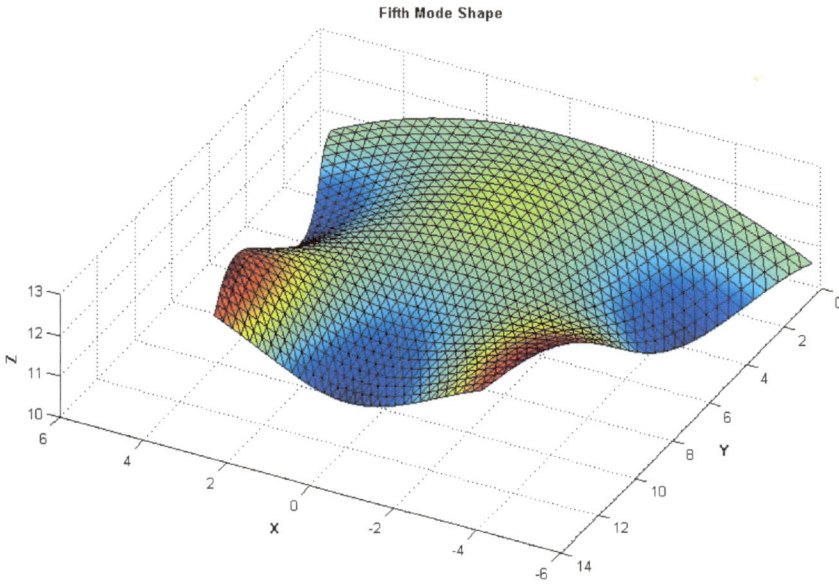

Figure 4.3(e): Fifth mode shape of fan blade (Symmetric mode).

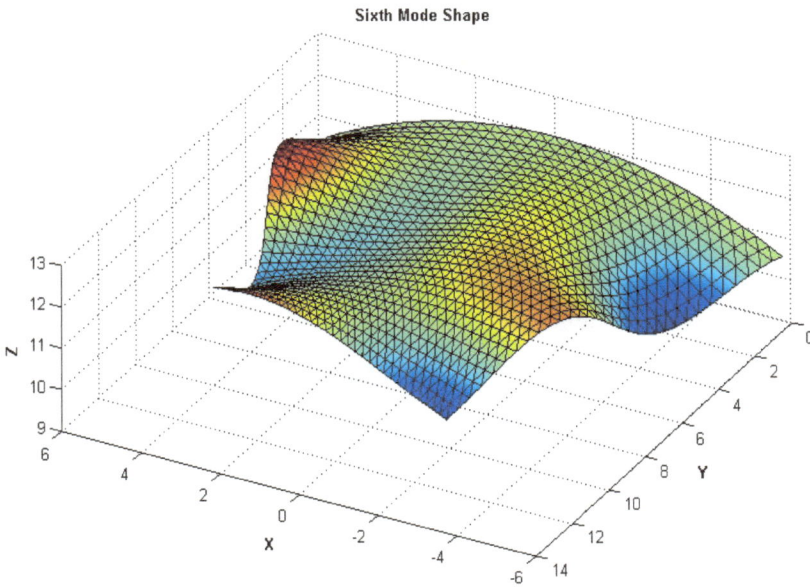

Figure 4.3(f): Sixth mode shape of fan blade (Anti-symmetric mode).

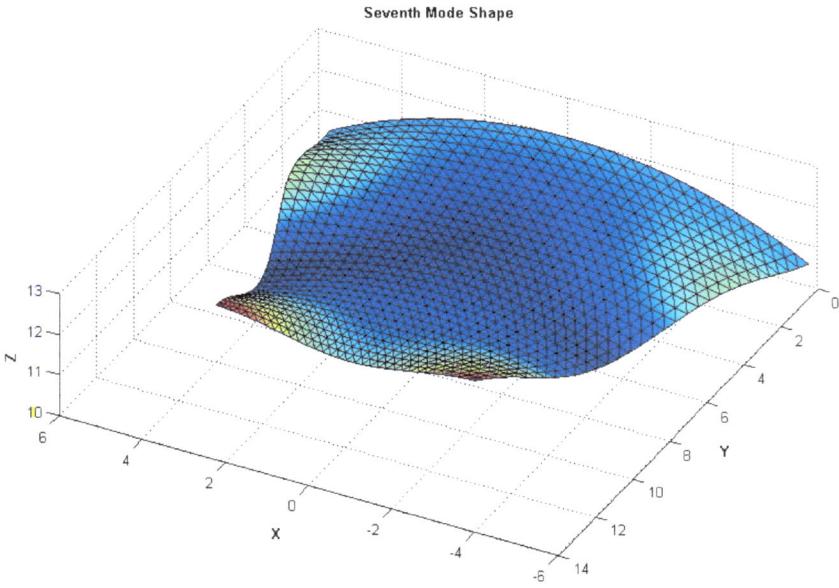

Figure 4.3(g): Seventh mode shape of fan blade (Symmetric mode).

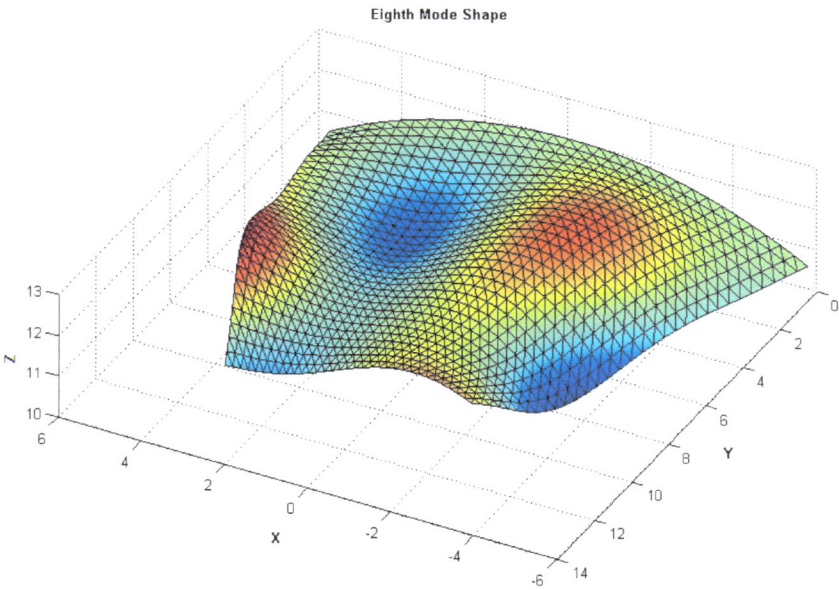

Figure 4.3(h): Eighth mode shape of fan blade (Anti-symmetric mode).

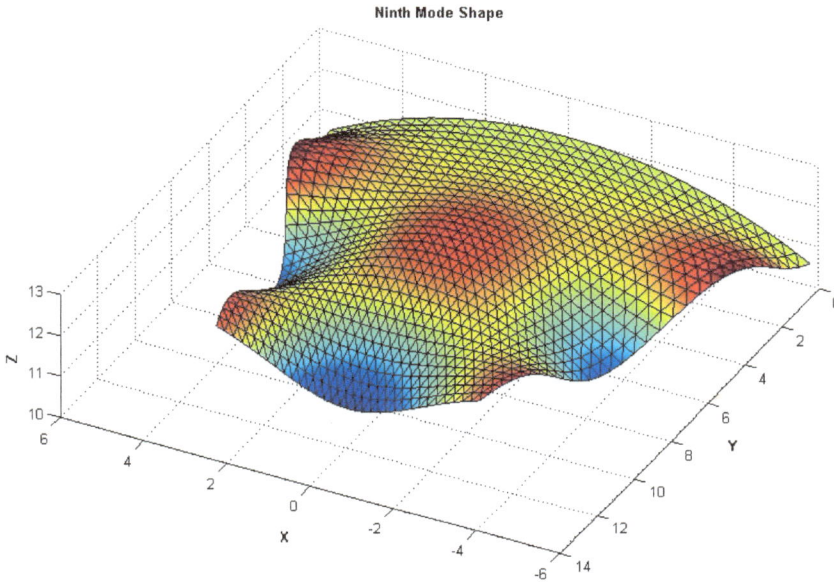

Figure 4.3(i): Ninth mode shape of fan blade (Symmetric mode).

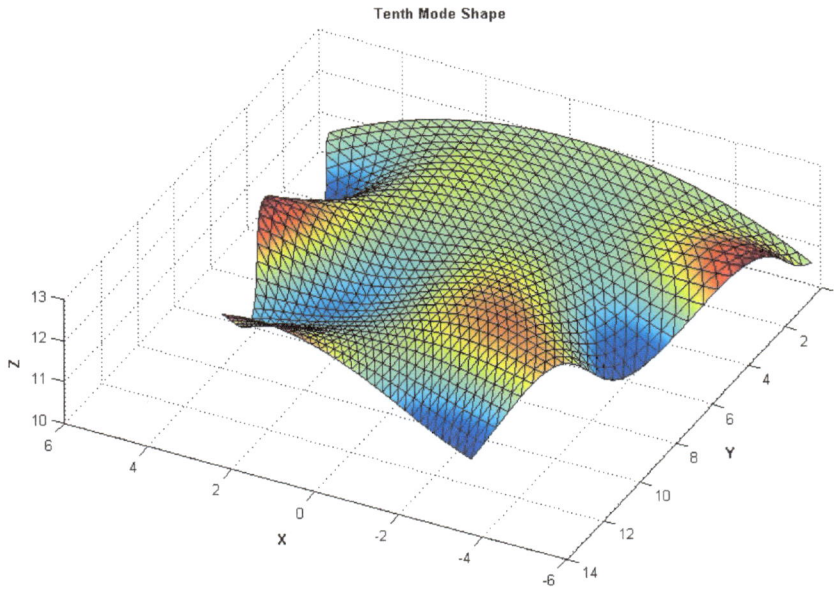

Figure 4.3(j): Tenth mode shape of fan blade (Anti-symmetric mode).

4.3. SCORDELIS-LO ROOF

This cylindrical roof, shown in Fig. **4.4**, is supported by rigid diaphragms at its two ends. The static analysis of this roof under uniform vertical load has been included in the obstacle course proposed by MacNeal and Harder [4.3]. For free vibration analysis, the pertinent data are: radius $R = 0.635$ m (25 in), length $L = 1.27$ m (50 in), thickness $h = 6.35$ mm (0.25 in), subtended angle or arc $B'AB = 80°$, modulus of elasticity $E = 2979$ GPa (4.32×10^8 psi which is from [4.3] and is very unrealistic), Poisson's ratio $v = 0$ and density $\rho_0 = 7739$ kg/m^3 (0.000725 lb·sec^2/in^4). A quarter of the roof is represented by the shell element introduced in Chapter 2 to determine the doubly symmetric natural frequencies and mode shapes. The boundary conditions are: $U = W = 0$ at curved edge AB, free at side BC, $V = \Theta_X = \Theta_Z = 0$ at CD, and $U = \Theta_Y = \Theta_z = 0$ at DA. Computed first ten frequency parameters are presented Table **4.3**. The first five doubly symmetric modes are included in Fig. **4.5** in which they are not plotted in scale. It may be appropriate to mention that the first three natural frequencies and mode shapes were previously investigated and reported in [4.4].

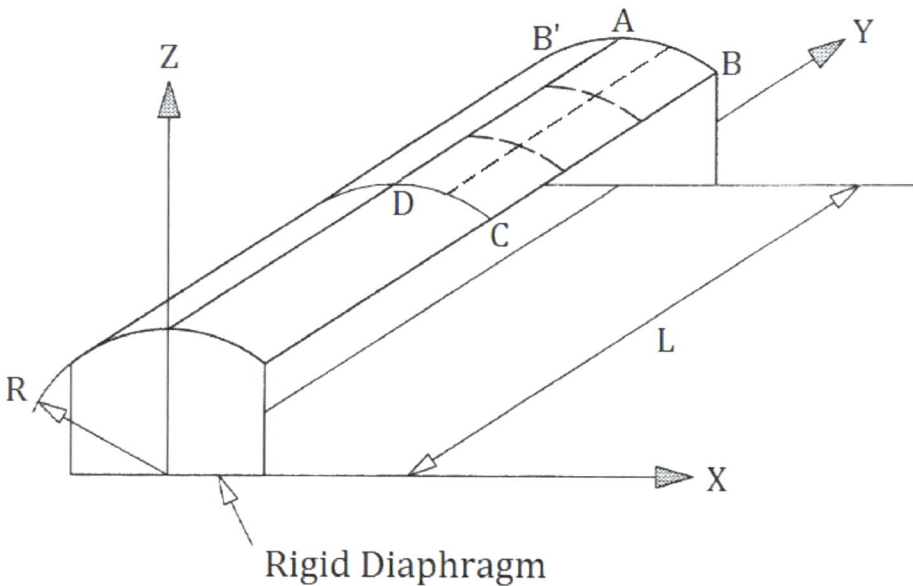

Figure 4.4: Geometrical configuration of Scordelis-Lo roof.

Table 4.3: First ten frequency parameters $\Omega = \omega R \sqrt{\frac{\rho_0 (1-\nu^2)}{E}}$ of Scordelis-Lo roof.

Mesh	NEQ	Ω_1	Ω_2	Ω_3	Ω_4	Ω_5
4×6D	306	0.0457143	0.173850	0.181047	0.205769	0.359371
8×10D	990	0.0451272	0.172959	0.178589	0.193500	0.350384
10×14D	1722	0.0450646	0.172787	0.177679	0.192483	0.346227
14×20D	3420	0.0450168	0.172664	0.177079	0.191640	0.343920
18×27D	5913	0.0449955	0.172613	0.176812	0.191326	0.342930
20×30D	7290	0.0449890	0.172599	0.176738	0.191237	0.342685
22×33D	8811	0.0449839	0.172589	0.176682	0.191173	0.342503
Mesh	**NEQ**	Ω_6	Ω_7	Ω_8	Ω_9	Ω_{10}
4×6D	306	0.371416	0.542001	0.600143	0.633677	0.653872
8×10D	990	0.358527	0.464264	0.548448	0.568925	0.629236
10×14D	1722	0.356949	0.457213	0.541558	0.556561	0.624641
14×20D	3420	0.355776	0.451357	0.535267	0.550270	0.621963
18×27D	5913	0.355293	0.449166	0.532841	0.547641	0.620854
20×30D	7290	0.355159	0.448554	0.532142	0.547034	0.620571
22×33D	8811	0.355060	0.448116	0.531634	0.546590	0.620363

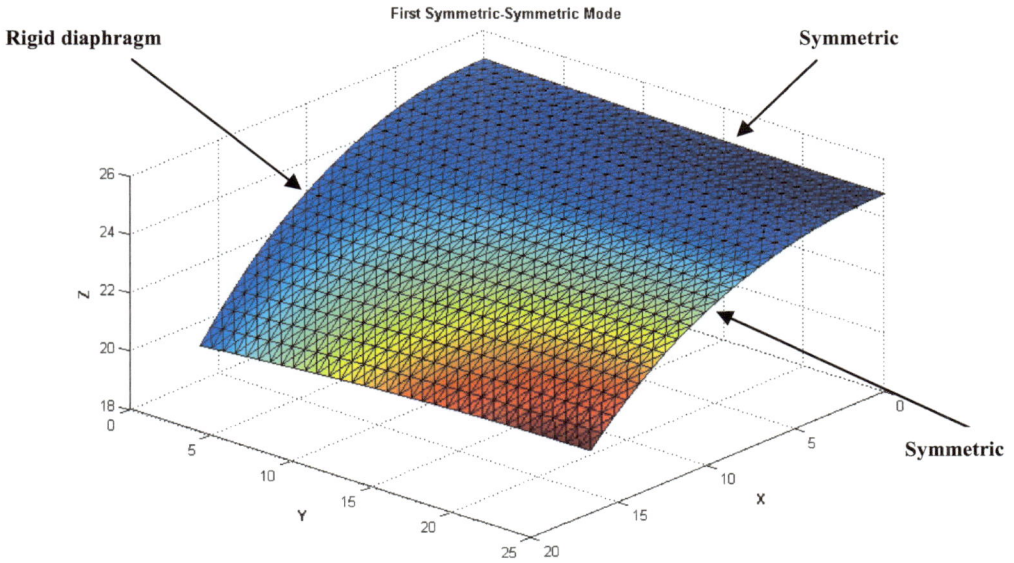

Figure 4.5(a): First symmetric-symmetric mode shape of Scordelis-Lo roof.

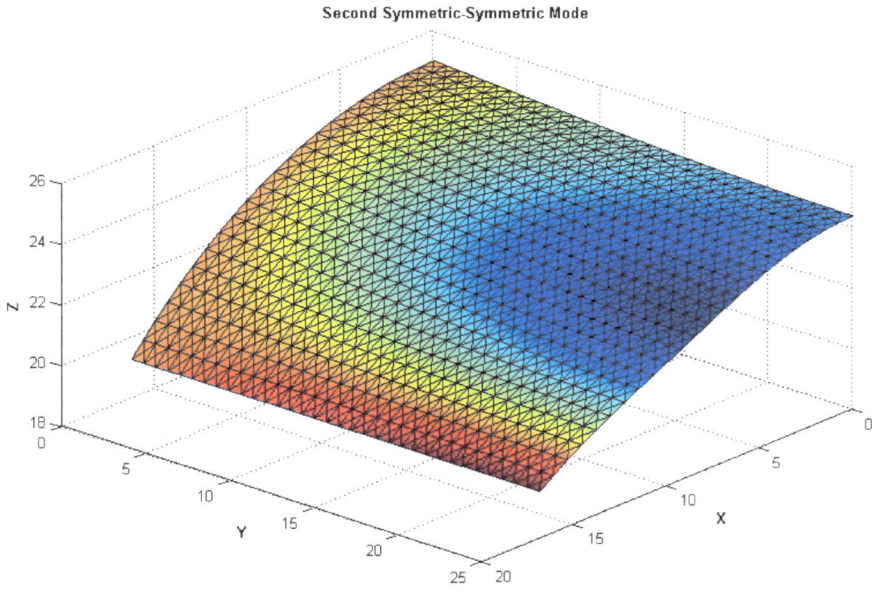

Figure 4.5(b): Second symmetric-symmetric mode shape of Scordelis-Lo roof.

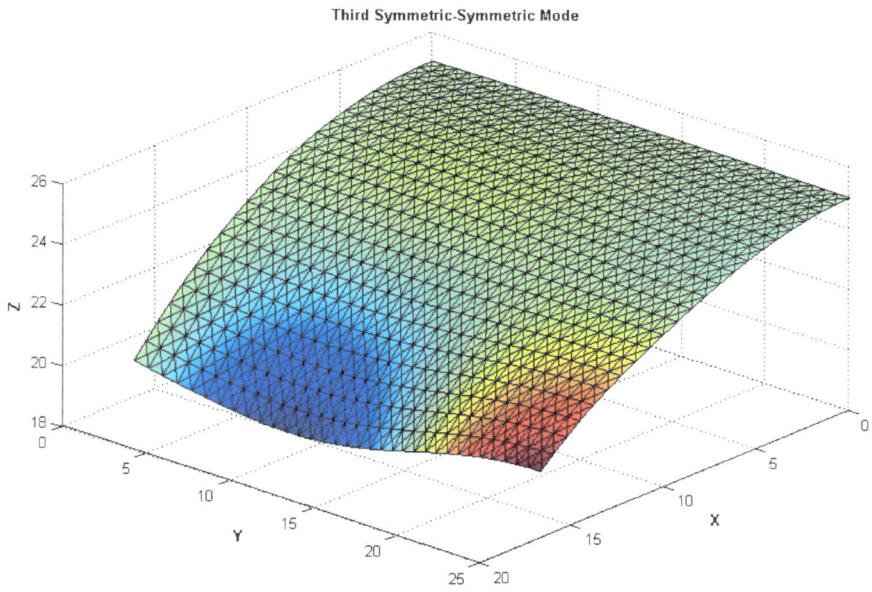

Figure 4.5(c): Third symmetric-symmetric mode shape of Scordelis-Lo roof.

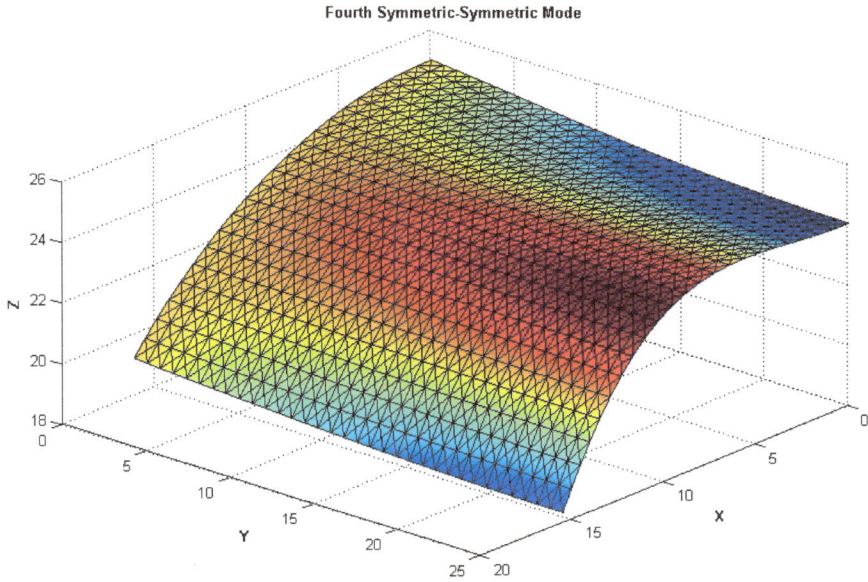

Figure 4.5(d): Fourth symmetric-symmetric mode shape of Scordelis-Lo roof.

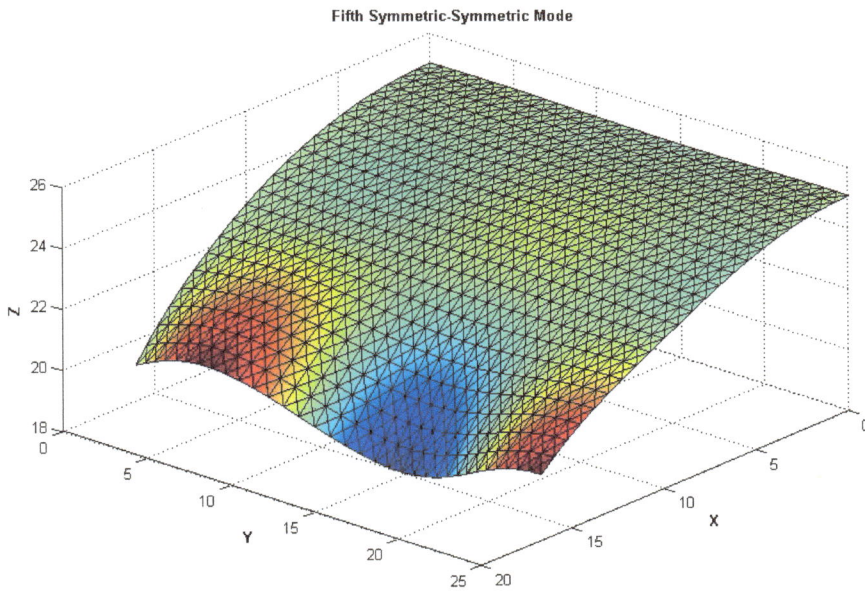

Figure 4.5(e): Fifth symmetric-symmetric mode shape of Scordelis-Lo roof.

4.4. CYLINDRICAL SHELL CLAMPED AT BOTH ENDS

In this section a circular cylindrical shell is studied. As shown in Fig. **4.6**, both ends of the shell are clamped. The shell has the following geometrical properties: $R = 76.2$ mm (3 in), $L = 304.8$ mm (12 in), and $h = 0.254$ mm (0.01in). Its material is a carbon steel with $E = 207$ GPa (30×10^6 psi), $v = 0.3$ and $\rho_0 = 7814$ kg/m^3 (0.000732 lb·sec^2/in^4). For finite element computations, one quarter of the shell (*ABCD* in Fig. **4.6**) is discretized. Combinations of symmetry and anti-symmetry conditions are subsequently applied to nodes along lines *AB* and *DC*. For nodes on line *AB*, symmetry means, $U = \Theta_Y = \Theta_Z = 0$ and anti-symmetry has $V = W = \Theta_X = 0$. These conditions become, for nodes on line *DC*, $V = \Theta_X = \Theta_Z = 0$ with symmetry and $U = W = \Theta_Y = 0$ with anti-symmetry. Nodes on arcs *AD* and *BC* have all six nodal DOF constrained to zero. Computed natural frequencies are given in Table **4.4**. Mesh designation is such that the first integer indicates the number of divisions over arc *AD* and the second integer is the number of divisions along *AB*. Examining the convergence of computed natural frequencies, it is found that fine meshes are required to obtain accurate results. For example, relative to the 14×34D mesh with NEQ = 5629, the 10×24D mesh having NEQ = 2820 yields discrepancies of 6.123% and 2.579%, on the first symmetric (circumferential or radial-wise) mode and the first anti-symmetric (circumferential or radial-wise) mode, respectively. Mesh 12×29D, having a NEQ of 4104, reduces such discrepancies to less than 2%. The need for fine meshes is due to the deep shell nature of the problem.

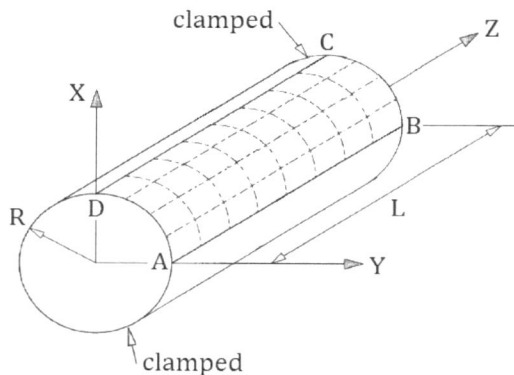

Figure 4.6: Fully clamped cylindrical shell.

Table 4.4: First ten natural frequencies (in Hz) of clamped cylindrical shell.

Mesh	NEQ	f_1 (1,6)[1]	f_2 (1,5)	f_3 (1,7)	f_4 (1,4)	f_5 (1,8)
5×12D	690	858.748	844.310	1196.15	1083.51	1384.02
8×19D	1776	644.067	628.954	784.333	787.313	1007.87
10×24D	2820	584.830	602.386	682.792	777.717	856.216
12×29D	4104	561.727	592.065	642.325	773.545	794.969
14×34D	5629	551.085	587.241	623.460	766.066	771.364
Mesh	NEQ	f_6 (2,7)[1]	f_7 (2,8)	f_8 (1,9)	f_9 (2,6)	f_{10} (2,9)
5×12D	690	1454.81	1853.08	1494.33	1959.45	1605.29
8×19D	1776	1047.12	1097.54	1174.19	1152.26	1295.72
10×24D	2820	973.454	1025.10	1083.23	1060.87	1169.07
12×29D	4104	944.413	974.425	996.713	1045.96	1104.90
14×34D	5629	930.706	950.522	955.477	1038.64	1067.68

[1]The integers inside the parentheses are (*m, n*) where *m* (which is not to be confused with the consistent element mass matrix in Chapter 2) the number of half-waves in the longitudinal direction, and *n* is the number of circumferential or radial waves, respectively.

The first ten mode shapes are presented in Fig. **4.7**. These mode shapes are useful in identifying mode numbers *m* and *n,* both being included in Table **4.4**. The sequencing of natural frequencies in the latter table is based on computational results associated with the 14×34D mesh. In Table **4.5** the computed natural frequencies with the latter mesh are included for comparison to available results in [4.5-4.6]. It should be mentioned that the results from the latter references were concerned with material properties: $E = 29.6 \times 10^6$ psi, $v = 0.29$ and $\rho_0 = 0.000733$ lb·sec^2/in^4. Thus, the computed results with the 14×34D mesh included in Table **4.5** are obtained by using the dimensionless frequency parameter $\Omega = \omega R \sqrt{\rho_0(1 - \nu^2)/E}$. It is observed that the present results are in general, higher than those from [4.5] and [4.6]. The smallest discrepancy, at 0.1724% with respect to the analytical results of [4.5], occurs at mode (1,5). Mode (1,4) has an 8.286% difference relative to the experimental result of [4.6]. Mode (1,5) is an anti-symmetric circumferential or radial mode while mode (1,4) is a symmetric circumferential or radial mode. Finally, it should be noted that the analytical results of [4.5] were obtained using the Flügge's shell theory.

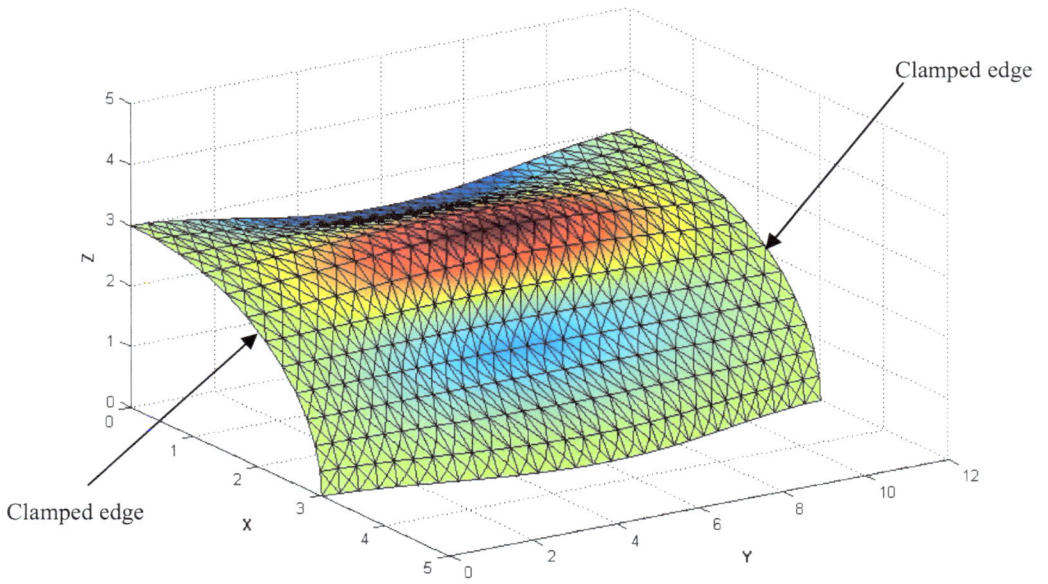

Figure 4.7(a): First mode shape of clamped cylinder [551.085 Hz, (1,6)].

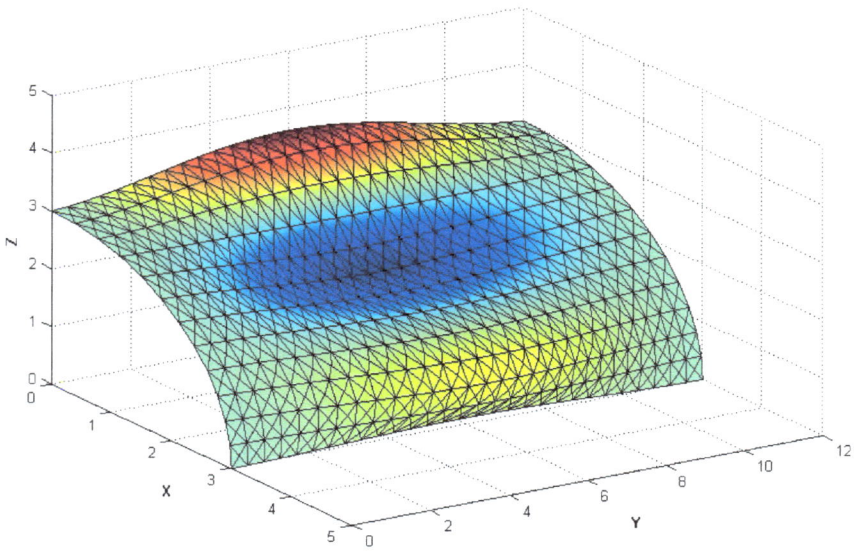

Figure 4.7(b): Second mode shape of clamped cylinder [587.241 Hz, (1,5)].

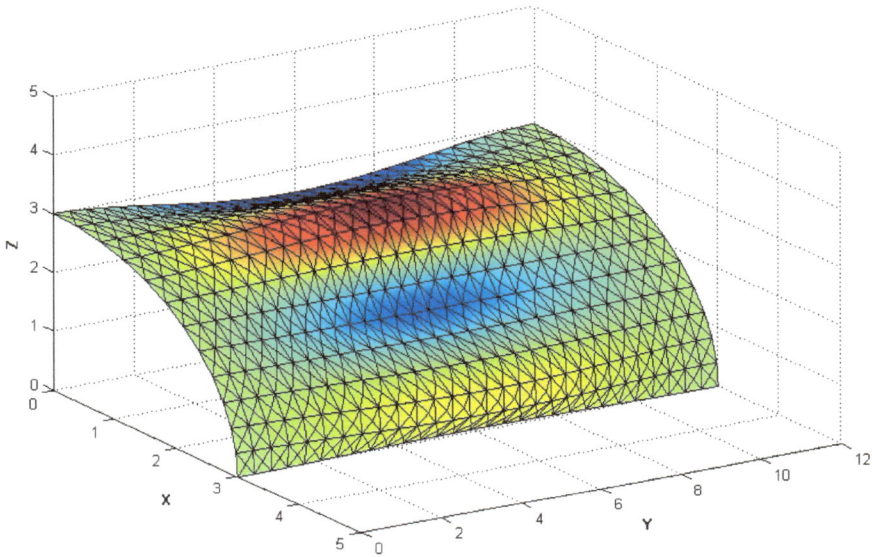

Figure 4.7(c): Third mode shape of clamped cylinder [623.460 Hz, (1,7)].

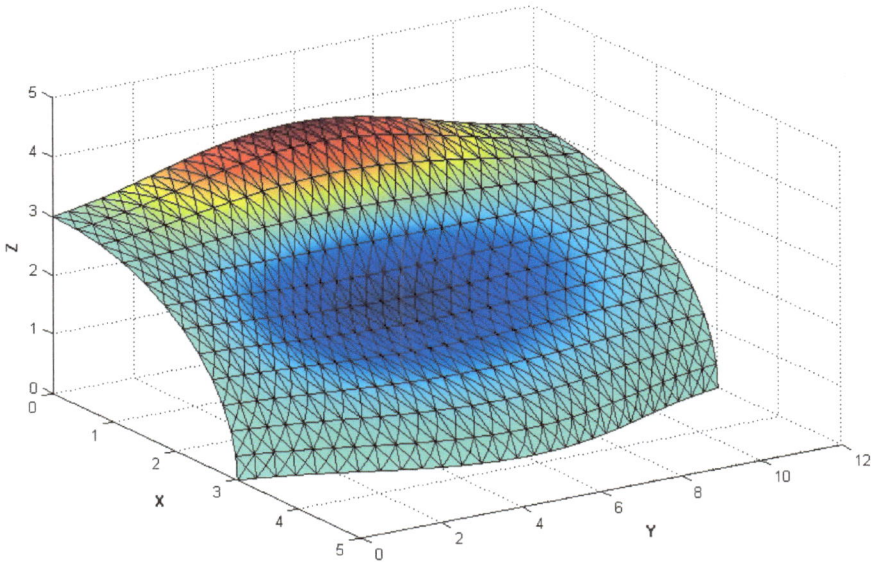

Figure 4.7(d): Fourth mode shape of clamped cylinder [766.066 Hz, (1,4)].

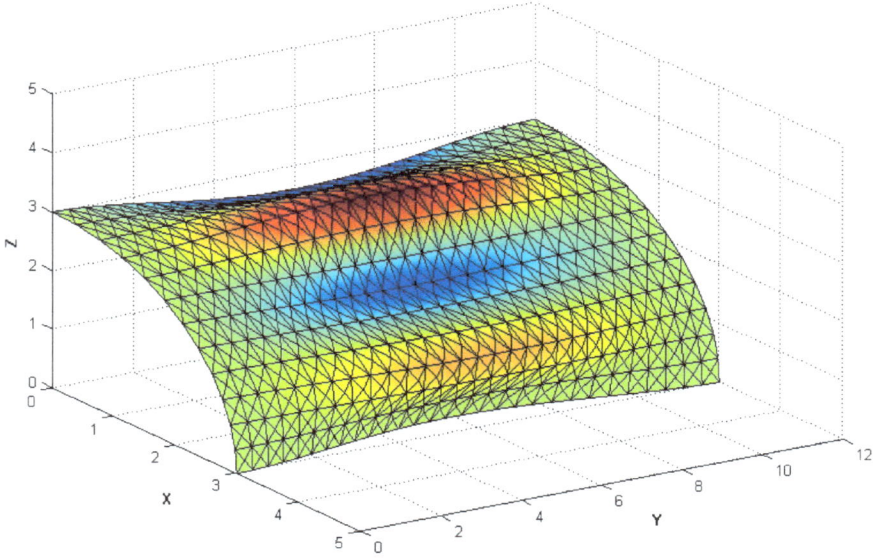

Figure 4.7(e): Fifth mode shape of clamped cylinder [771.364 Hz, (1,8)].

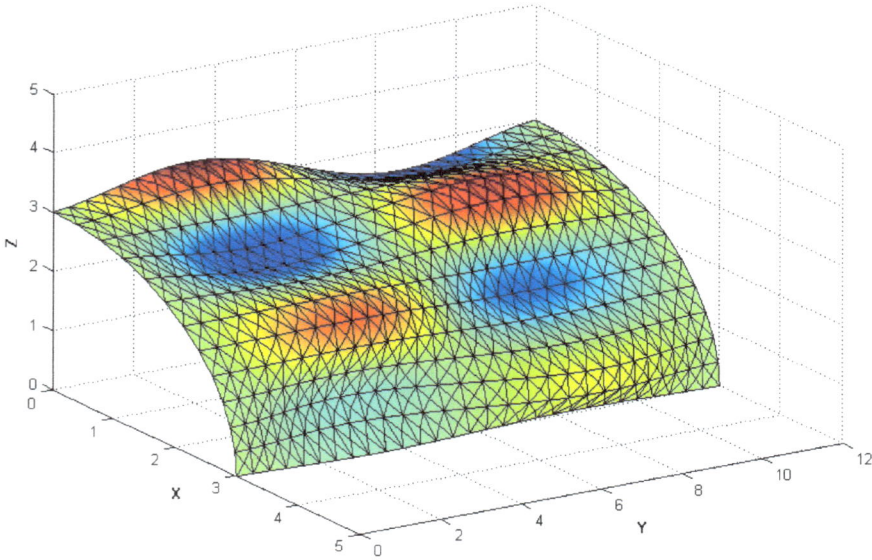

Figure 4.7(f): Sixth mode shape of clamped cylinder [930.706 Hz, (2,7)].

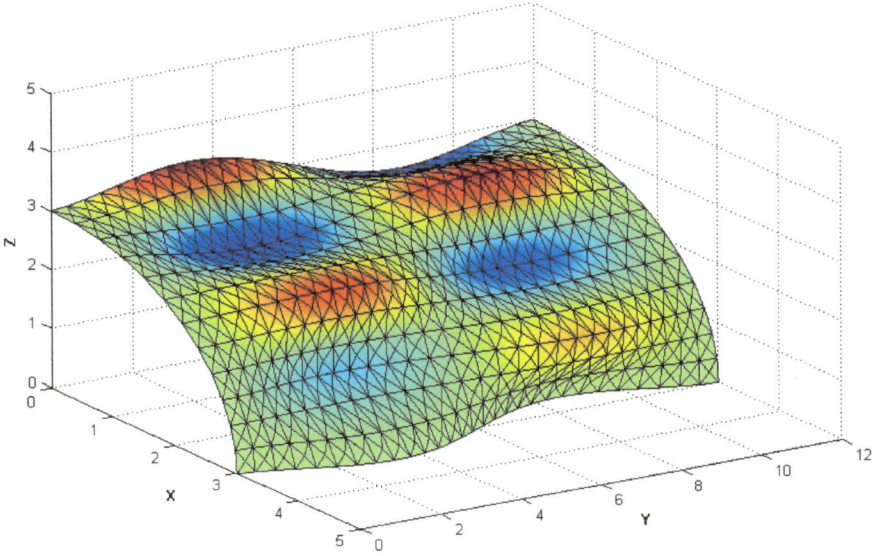

Figure 4.7(g): Seventh mode shape of clamped cylinder [950.522 Hz, (2,8)].

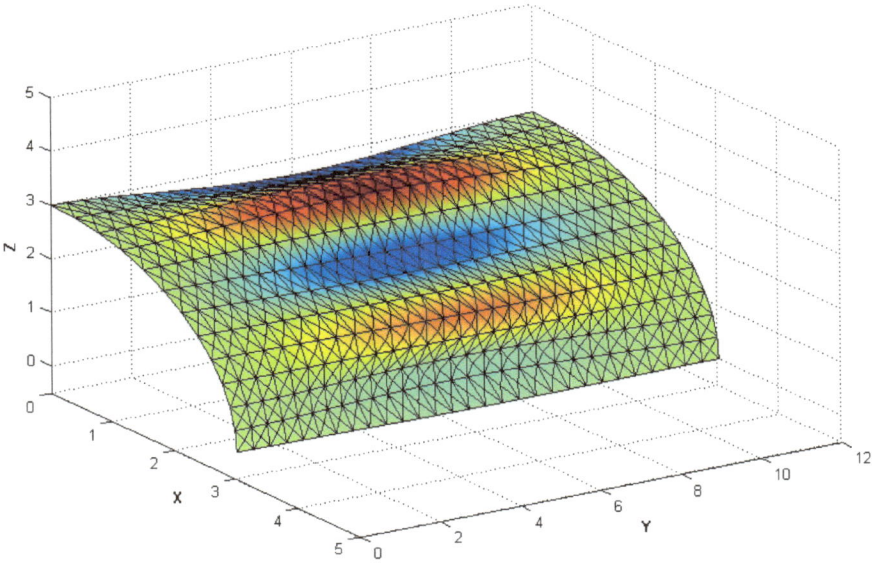

Figure 4.7(h): Eighth mode shape of clamped cylinder [955.477 Hz, (1,9)].

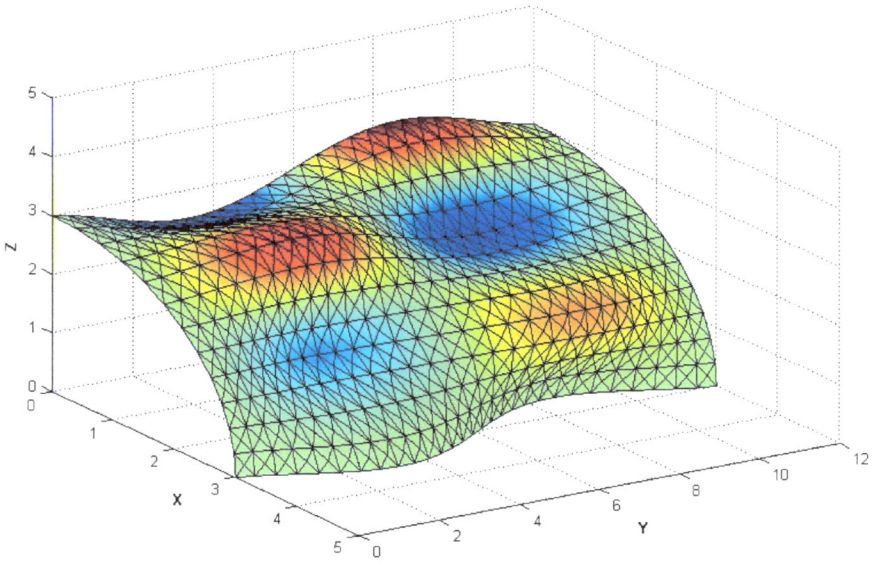

Figure 4.7(i): Ninth mode shape of clamped cylinder [1038.64 Hz, (2,6)].

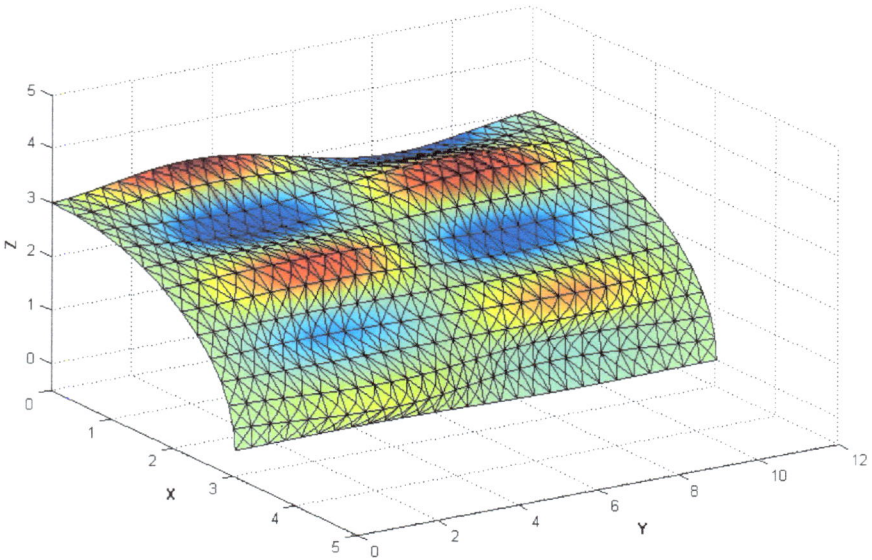

Figure 4.7(j): Tenth mode shape of clamped cylinder [1067.68 Hz, (2,9)].

Table 4.5: Comparison of lower natural frequencies (in Hz) of clamped cylindrical shell.

Method	Mode numbers (*m, n*)						
	(1,6)	**(1,5)**	**(1,7)**	**(1,4)**	**(1,8)**	**(2,7)**	**(2,8)**
14×34D[1]	546.0	581.0	617.0	758.0	763.0	911.0	940.0
[4.5][2]	538.0	580.0	597.0	765.0	721.0	907.0	911.0
[4.6][3]	525.0	559.0	587.0	700.0	720.0	875.0	900.0

[1] Results with due modification to material properties for comparison.
[2] Analytical results.
[3] Experimental results.

4.5. EFFECT OF ASPECT RATIO FOR CYLINDRICAL SHELL CLAMPED AT BOTH ENDS

The effects of the aspect ratio (radius-to-thickness ratio) for the cylindrical shell structure in Fig. **4.6** are studied in this section. However, the present shell structure has the following properties: $R = 76.2$ mm (3 in), $L = 304.8$ mm (12 in), $E = 203$ GPa (29.6×10^6 psi), $v = 0.3$ and $\rho_0 = 7825$ kg/m^3 (0.000733 lb·sec^2/in^4). The thickness is given three values, 0.254 mm (0.01 in), 1.27 mm (0.05 in) and 2.54 mm (0.1 in). Again the *ABCD* quarter (see Fig. **4.6**) is discretized for the computation. To limit the scope of the study, only symmetric boundary conditions are applied to edges *AB* and *DC*. The computed symmetric modes are included in Table **4.6** where they are compared with the analytical results of [4.5]. It is interesting to note the following.

(i)   The finding of upper bound solutions in the thin regime and lower bound solutions in the moderately-thick to thick regime (see Sec. 3.2) holds true.

(ii) The transition point of upper bound solutions to lower bound solutions is at aspect ratio of 30. This is consistent with the finding for the clamped circular plate investigated in Sec. 3.2.

(iii) A fine (finer than 14×34D, or NEQ ≥ 5629) mesh would be necessary to provide the accurate vibration characteristics of a cylindrical shell with radius-to-thickness ratio of 10 or less.

Table 4.6: Comparison of effects of aspect ratio on natural frequencies (in Hz) of clamped cylindrical shell.

Method	$R/h = 100$		$R/h = 20$		$R/h = 10$	
	(1,6)[1]	(1,4)	(1,4)	(1,2)	(1,4)	(1,2)
14×34D[2]	546.0	758.0	1113.0	1938.0	1798.0	1971.0
[4.5][3]	538.0	765.0	1148.0	1940.0	1895.0	1982.0
difference (%)	1.487	-0.9150	-3.049	-0.1031	-5.119	-0.5500

[1] Mode numbers (m, n).
[2] NEQ = 5629.
[3] Analytical results.

REFERENCES

[4.1] S.C. Fan, and M.H. Luah, "Free vibration analysis of arbitrary thin shell structures by using spline finite element", *J. Sound Vibr.*, vol. 179, pp. 763-776, February 1995.

[4.2] H.T.Y. Yang, and Y.C. Wu, "A geometrically nonlinear tensorial formulation of a skewed quadrilateral thin shell finite element", *Int. J. Numer. Methods Engrg.*, vol. 28, pp. 2855-2875, December 1989.

[4.3] R.H. MacNeal, and R.L. Harder, "A proposed standard set of problems to test finite element accuracy", *J. Finite Elem. Anal. Des.*, vol. 1, pp. 3-20, April 1985.

[4.4] M.L. Liu, and C.W.S. To, "Vibration of structures by hybrid strain-based three-node flat triangular shell elements", *J. Sound and Vibration,* vol. 184(5), pp. 801-821, 1995.

[4.5] B.L. Smith, and E.E. Haft, "Natural frequencies of clamped cylindrical shells", *AIAA J.*, vol. 6, pp. 720-721, April 1968.

[4.6] L.R. Koval, and E.T. Cranch, "On the free vibrations of thin cylindrical shells subjected to an initial torque", in Proceedings of the 4th U.S. National Congress in Applied Mechanics, 1962, pp. 107-117.

Send Orders for Reprints to reprints@benthamscience.net
Vibration and Nonlinear Dynamics of Plates and Shells, 2014, 77-99   **77**

CHAPTER 5

Vibration Analysis of Shells with Double Curvatures

Abstract: Vibration characteristics of shell structures with double curvatures are studied in this chapter. These shell structures include the spherical caps, spherical panel of square projection, hemispherical panel, and hemispherical shell. The computed results are compared with available data in the literature whenever they are available.

Keywords: Vibration analysis, spherical shells, caps, panel, hemispheres.

5.1. SPHERICAL CAPS AND EFFECT OF ASPECT RATIO

The shallow spherical caps shown in Fig. **5.1** are investigated in this section. First, the cap is simply supported and the effect of aspect ratio is examined. Next, the cap is clamped and its natural frequencies are computed. A quarter of the cap or the entire cap is approximated by the triangular shell finite elements introduced in Chapter 2.

With reference to Fig. **5.1**, the simply supported cap has the following parameters or properties: $R = 1.0$ m, $\varphi = 14.5°$, $a = 0.5$ m, $E = 200$ GPa, $v = 0.3$ and $\rho_0 = 7830$ kg/m^3. The thickness is set to five different values: $h = 0.001$ m, 0.1 m, 0.2 m, 0.3 m and 0.4 m, respectively. A quarter of the cap is discretized, using three different meshes, 297D which is shown in Fig. **3.3(c)**, and 444D and 810D which are given in Fig. **5.2**. The sides of $X = 0$ and $Y = 0$ are symmetry lines. Thus, for the side $X = 0$, $U = \Theta_Y = \Theta_Z = 0$ while the side $Y = 0$, $V = \Theta_X = \Theta_Z = 0$. For the curved edge where the cap is simply supported, the boundary conditions are: $U = V = W = 0$. Consequently, only the symmetric-symmetric modes are obtained. The computed frequency parameters are listed in Table **5.1**, and compared with those from [5.1]. Some comments are in order.

(i)   The Ritz method was used by [5.1], and the results are hence upper bound solutions.

(ii)   The present element yields higher fundamental frequency when $R/h = 2000$, but lower fundamental frequency for other values of R/h.

(iii) The transition point from lower bound to upper bound for the presently obtained results cannot be established since it is a situation of a upper or lower bound solution being compared with a upper bound solution.

(iv) The observation in Sec. 3.2 that the thicker the shell, the higher the discrepancy, in terms of absolute values, holds true. In other words, the thicker the shell, the finer the mesh is required to obtain accurate natural frequencies.

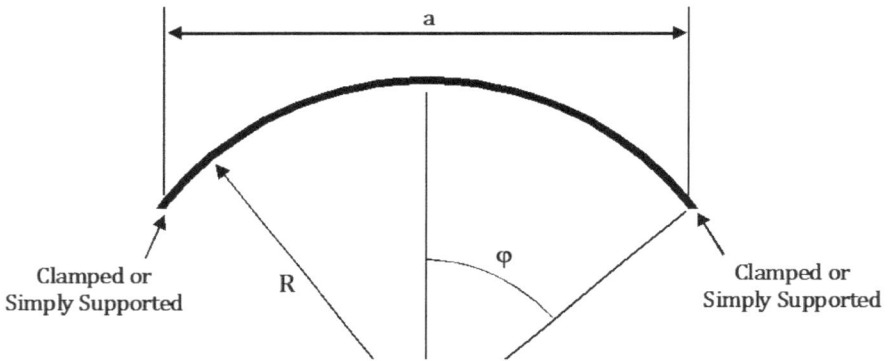

Figure 5.1: Clamped or simply-supported spherical cap.

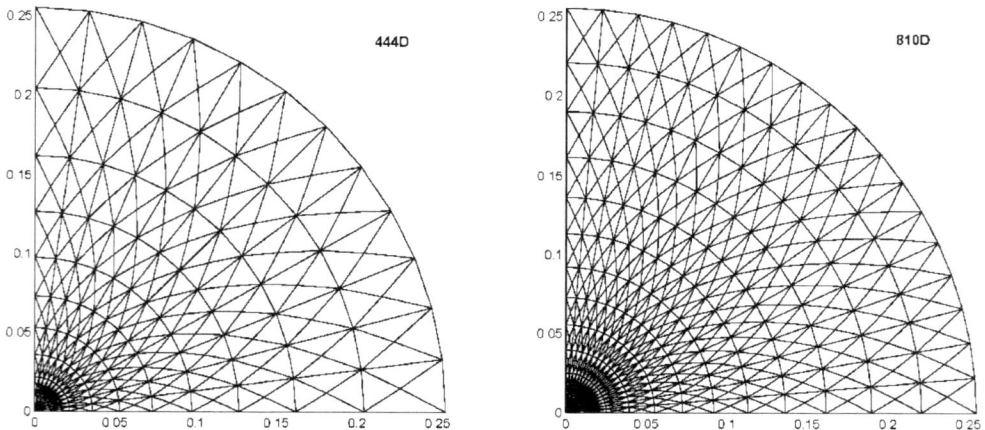

Figure 5.2: Meshes 444D and 810D.

Table 5.1: Comparison of frequency parameter $\Omega_1 = \omega_1 a \sqrt{\frac{\rho_0}{E}}$ of spherical cap (simply supported).

	$R/h = 2000$	$R/h = 20$	$R/h = 10$	$R/h = 2/3$	$R/h = 5$
[5.1]	0.49961	0.95008	1.2999	1.6150	1.8623
297D[1]	0.504125	0.920389	1.21086	1.45624	1.63534
difference (%)	0.9047	-3.125	-6.850	-9.830	-12.19
444D[2]	0.503032	0.921124	1.21753	1.47306	1.66317
difference (%)	0.6849	-3.048	-6.272	-8.789	-10.69
810D[3]	0.501742	0.921674	1.22341	1.48814	1.68843
difference (%)	0.4268	-2.990	-5.884	-7.855	-9.336

[1] NEQ = 891.
[2] NEQ = 1332.
[3] NEQ = 2430.

The second part of this section is concerned with natural frequencies of two different clamped spherical caps. The first cap has the same geometrical and material properties as the simply supported cap above, with the exception that the thickness of the cap is set to be $h = 0.001$ m only, resulting in a thin spherical shell. A quarter of the cap is represented by mesh 444D. All DOF associated with the nodes at the curved edge are constrained to zero. In order to capture all vibration modes, be them symmetric (about X-axis)-symmetric (about Y-axis) (SS), anti-symmetric-anti-symmetric (AA), symmetric-anti-symmetric (SA), and anti-symmetric-symmetric (AS), the sides of $X = 0$ and $Y = 0$ are applied, respectively, symmetric and symmetric, anti-symmetric and symmetric, and anti-symmetric and anti-symmetric boundary conditions. Symmetric and anti-symmetric boundary conditions are also applied. SA and AS modes are identical in their frequencies. Their mode shapes are 90° different; that is, rotating an AS mode shape by 90° will result in the corresponding SA mode shape. For brevity, only SS, AA and AS modes will be presented.

Symmetry boundary conditions are: $U = \Theta_Y = \Theta_Z = 0$ for side $X = 0$, and $V = \Theta_X = \Theta_Z = 0$ for side $Y = 0$. Anti-symmetry conditions are: $V = W = \Theta_X = 0$ for side $X = 0$, and $U = W = \Theta_Y = 0$ for side $Y = 0$. The computed frequency parameters are presented in Tables **5.2** through **5.4**. The first three frequency parameters in every mode category are also compared with those obtained by the Ritz method [5.1]. Excellent agreements are observed in all categories. Note that the AA modes can be captured by applying the SS boundary conditions. For example, AA-1 and SS-2 have the same frequency value, and their mode shapes are 45° apart from each other, see Fig. **5.3**.

Table 5.2: First five SS frequency parameters $\Omega = \omega a \sqrt{\frac{\rho_0}{E}}$ of clamped spherical cap.

	Modes				
	1	2	3	4	5
444D[1]	0.501907	0.509175	0.512944	0.520214	0.531982
[5.1]	0.50048	0.50616	0.50979	N/A	N/A
difference (%)	0.2851	0.5957	0.6187	N/A	N/A

[1] NEQ = 1297; N/A denotes "not available".

Table 5.3: First five AS frequency parameters $\Omega = \omega a \sqrt{\frac{\rho_0}{E}}$ of clamped spherical cap.

	Modes				
	1	2	3	4	5
444D[1]	0.504809	0.514272	0.520910	0.526649	0.541835
[5.1]	0.50293	0.51040	0.51655	N/A	N/A
difference (%)	0.3736	0.7586	0.8441	N/A	N/A

[1] NEQ = 1298.

Table 5.4: First five AA frequency parameters $\Omega = \omega a \sqrt{\frac{\rho_0}{E}}$ of clamped spherical cap.

	Modes				
	1	2	3	4	5
444D[1]	0.509175	0.520214	0.531982	0.534317	0.551445
[5.1]	0.50616	0.51564	0.52499	N/A	N/A
difference (%)	0.5957	0.8871	1.332	N/A	N/A

[1] NEQ = 1297.

The second clamped spherical cap has, $R = 1.0$ m, $\varphi = 10°$, $a = 0.35$ m, $h = 0.005$ m, $E = 200$ GPa, $v = 0.3$ and $\rho_0 = 7830$ kg/m^3. The entire cap is discretized. A sample mesh, 32×8D, is shown in Fig. **5.4**. The two integers in the mesh designator indicate, respectively, the numbers of divisions along the longitudinal direction and latitudinal direction. In Table **5.5**, the first 15 frequency parameters $\Omega = \omega \sqrt{\rho_0(1 - v^2)/E}$ that are computed for three meshes are listed. They are compared with results of Ref.

[5.2] which are given in the last row of the table. In Ref. [5.2] obtained the first three axi-symmetric modes were obtained by using eight-node degenerate isoparametric shell elements. The meshes in [5.2] have a small cutout of $0.01°$ around the pole. The convergence study (see Table **5.1** in [5.2]) indicated that the three axi-symmetric natural frequencies were lower bound solutions. By examining the presently computed mode shapes, it is found that the first, sixth and fifteen modes correspond to axi-symmetric modes. The discrepancies are, comparing 40×10D with [5.2], 5.113%, 4.399% and 4.464%, suggesting finer mesh if better accuracy is desired. Finally, for brevity only the three axi-symmetric mode shapes are plotted in Fig. **5.5** in which the scale of Z-axis is highly exaggerated for visual effect.

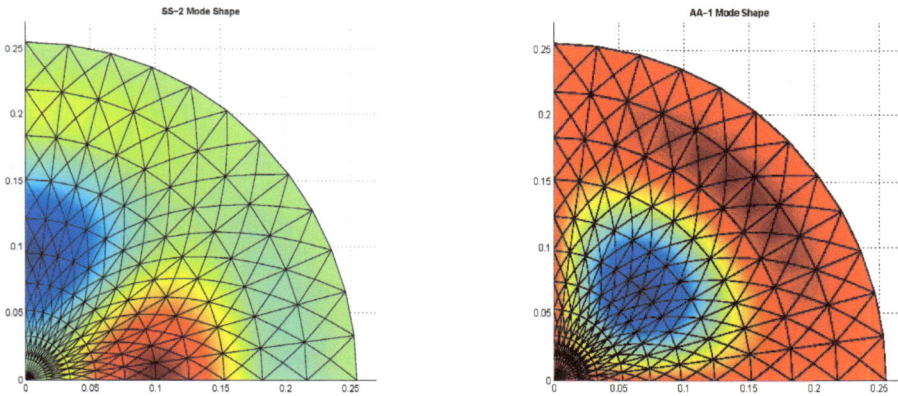

Figure 5.3: Mode shapes SS-2 and AA-1 of clamped spherical cap.

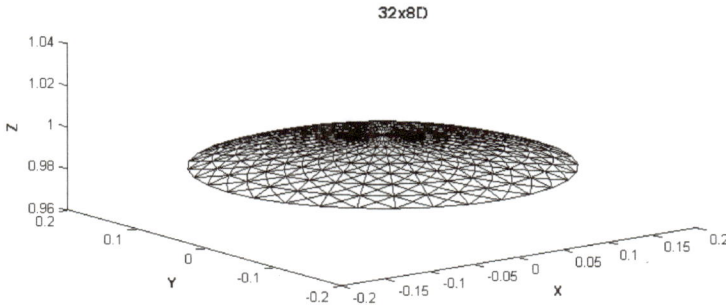

Figure 5.4: Sample discretization of entire spherical cap.

Table 5.5: First 15 frequency parameters $\Omega = \omega\sqrt{\frac{\rho_0(1-\nu^2)}{E}}$ of clamped spherical cap.

Mesh	NEQ	Ω_1	$\Omega_2 = \Omega_3$	$\Omega_4 = \Omega_5$	Ω_6	$\Omega_7 = \Omega_8$
20×5D	966	1.47743	1.53998	2.11008	2.40609	2.89691
32×8D	2694	1.46484	1.51402	2.04245	2.32286	2.76322
40×10D	4326	1.46173	1.50835	2.02887	2.30563	2.73657
[5.2]		1.3906	N/A	N/A	2.1875	N/A

Mesh	NEQ	$\Omega_9 = \Omega_{10}$	$\Omega_{11} = \Omega_{12}$	$\Omega_{13} = \Omega_{14}$	Ω_{15}
20×5D	966	3.43027	3.88177	4.74163	5.08127
32×8D	2694	3.22253	3.64872	4.35351	4.64953
40×10D	4326	3.18195	3.60166	4.28709	4.56143
[5.2]		N/A	N/A	N/A	4.3665

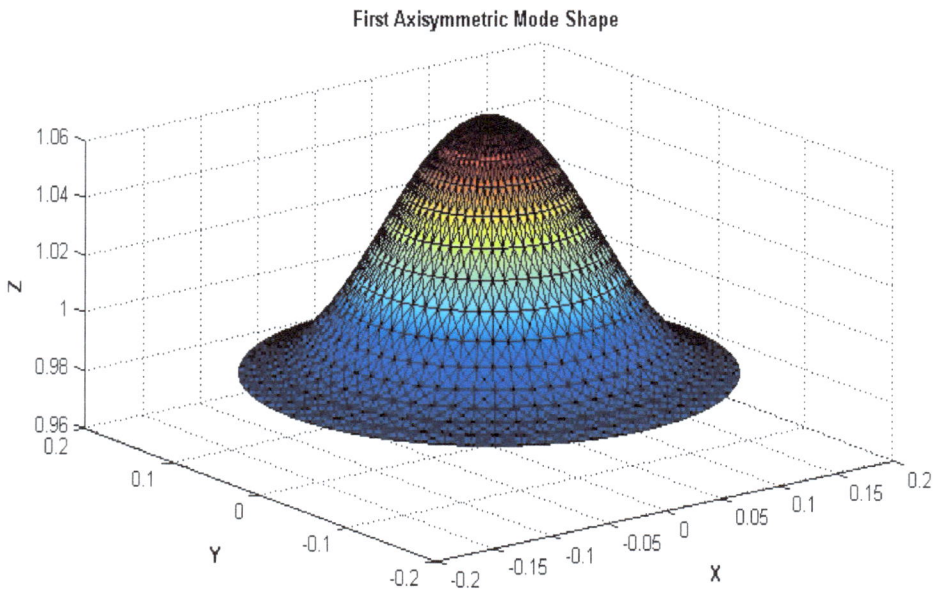

Figure 5.5(a): First axi-symmetric mode shape of spherical cap.

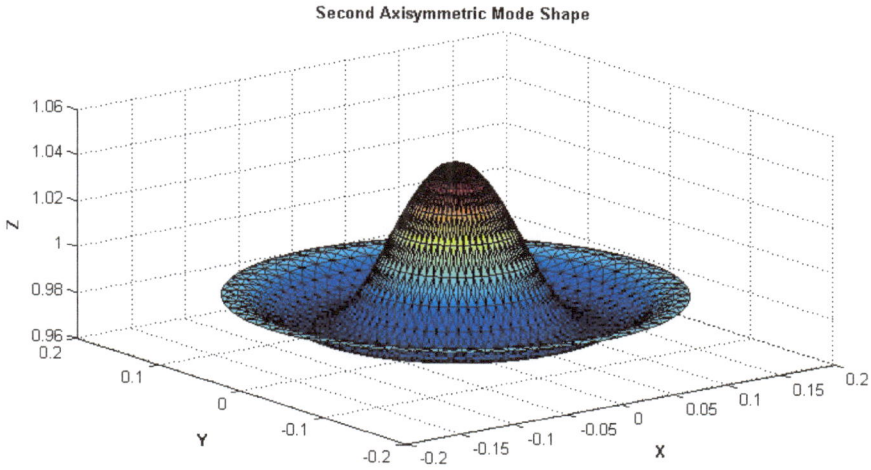

Figure 5.5(b): Second axi-symmetric mode shape of spherical cap.

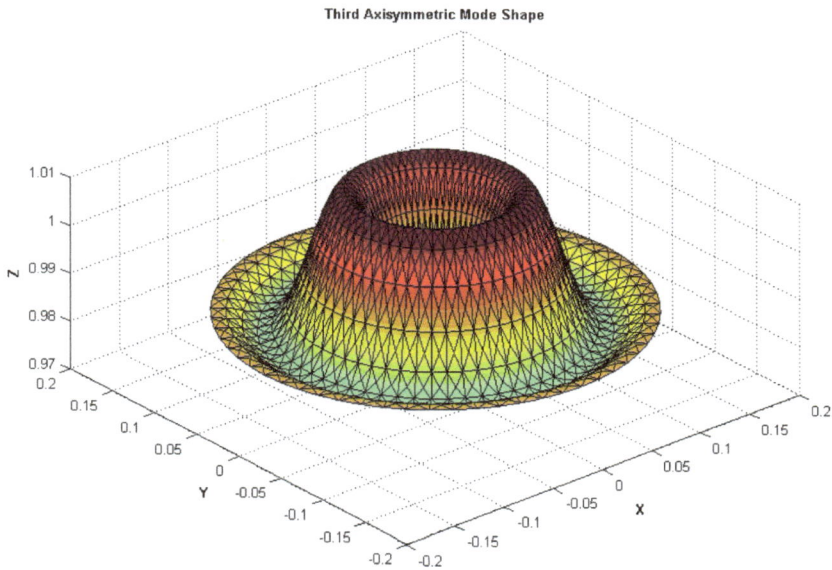

Figure 5.5(c): Third axi-symmetric mode shape of spherical cap.

5.2. SPHERICAL PANEL OF SQUARE PROJECTION

The spherical panel of square projection is shown in Fig. **5.6**. It is clamped at all edges. Geometrical dimensions are: $R = 1.0$ m, $a = b = 0.5$ m and $h = 0.05$ m. Material properties are: $E = 200$ GPa, $v = 0.3$ and $\rho_0 = 7830$ kg/m^3. In the finite element modeling, the entire panel is discretized. Nodes on the four edges are constrained. Computed frequency parameters $\Omega = wa\sqrt{\rho_0/E}$ for the first ten natural frequencies are presented in Table **5.6** for type C meshes and in Table **5.7** for type D meshes. In the mesh designation, the first and second integers indicate number of divisions along the *X*- and *Y*-directions, respectively. Compared with the two sets of results from [5.3, 5.4], the presently computed frequencies have very good agreements. It should be mentioned that the methods employed in [5.3, 5.4] are of Ritz-type. Specifically, [5.3] dealt with thin shell application and [5.4] with thick shell vibration analysis. In terms of mesh size, mesh 12×12C or 8×8D with NEQ of 726 and 678, respectively, is able to produce accurate (< 5% discrepancy) natural frequencies up to the fifth. On the other hand, mesh 16×16C or 12×12D, having a NEQ of 1350 and 1590, respectively, yields accurate results for all ten frequencies. These findings are consistent with those in previous sections (for example, Sec. 4.2 and Sec. 4.4). However, it is interesting to note that the D meshes are able to capture repeated natural frequencies whereas the C meshes fail to produce repeated frequencies, although in terms of mesh layout, both mesh types result in symmetric meshes.

Figure **5.7** shows the first and second, sixth and seventh, and ninth and tenth mode shapes computed with the 24×24D mesh. These pairs are the repeated natural frequencies but with mode shapes being related to each other. Specifically, for example, the pair can consist of one with a mode shape being rotated counter-clockwise by 90° to obtain the other mode shape with the same natural frequency.

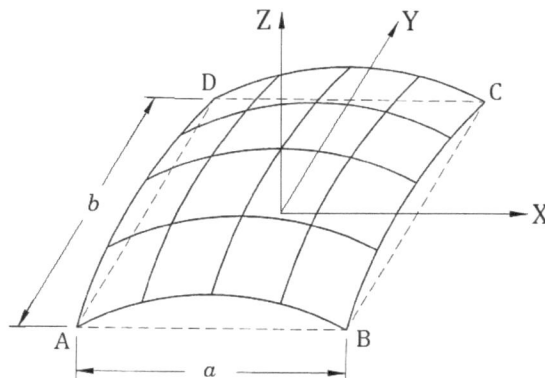

Figure 5.6: Spherical panel with square projection ($b = a$).

Table 5.6: Frequency parameters $\Omega = \omega a\sqrt{\frac{\rho_0}{E}}$ of clamped spherical panel ($R/h = 20$, $R/b = 2$, $b/a = 1$, type C meshes).

Mesh	NEQ	Ω_1	Ω_2	Ω_3	Ω_4	Ω_5
8×8C	294	0.589349	0.601228	0.632095	0.687312	0.720677
12×12C	726	0.574219	0.585796	0.597990	0.646711	0.668622
16×16C	1350	0.569124	0.580840	0.588641	0.635170	0.654164
20×20C	2166	0.566693	0.578600	0.584604	0.630326	0.648037
24×24C	3174	0.565321	0.577394	0.582445	0.627827	0.644869
28×28C	4374	0.564460	0.576668	0.581138	0.626366	0.643014
32×32C	5756	0.563879	0.576195	0.580282	0.625435	0.641833
[5.3][1]		0.58099	0.58099	0.59594	0.63537	0.65422
[5.4][2]		0.57638	0.57638	0.59134	0.63038	0.64764
Mesh	**NEQ**	Ω_6	Ω_7	Ω_8	Ω_9	Ω_{10}
8×8C	294	0.818417	0.894876	0.911477	1.01691	1.03232
12×12C	726	0.775819	0.778875	0.789858	0.867480	0.875063
16×16C	1350	0.745546	0.748538	0.781584	0.831343	0.841033
20×20C	2166	0.733640	0.736606	0.777461	0.816908	0.827547
24×24C	3174	0.727770	0.730738	0.775140	0.809634	0.820790
28×28C	4374	0.724450	0.727429	0.773708	0.805451	0.816926
32×32C	5756	0.722390	0.725383	0.772762	0.802826	0.814516
[5.3][1]		0.73299	0.73299	0.77902	N/A	N/A
[5.4][2]		0.72609	0.72609	0.77493	N/A	N/A

[1] Thin shell results,
[2] Thick shell results.

Table 5.7: Frequency parameters $\Omega = \omega a\sqrt{\frac{\rho_0}{E}}$ of clamped spherical panel ($R/h = 20$, $R/b = 2$, $b/a = 1$, type D meshes).

Mesh	NEQ	Ω_1	Ω_2	Ω_3	Ω_4	Ω_5
8×8D	678	0.585882	0.585882	0.613536	0.648487	0.681354
12×12D	1590	0.581522	0.581522	0.600849	0.639572	0.661640
16×16D	2886	0.579992	0.579992	0.596877	0.636523	0.655743
20×20D	4566	0.579279	0.579279	0.595159	0.635121	0.653223
24×24D	6630	0.578889	0.578889	0.594277	0.634362	0.651934
[5.3][1]		0.58099	0.58099	0.59594	0.63537	0.65422
[5.4][2]		0.57638	0.57638	0.59134	0.63038	0.64764

[1] Thin shell results.
[2] Thick shell results.

Table 5.7: (cont'd) Frequency parameters of clamped spherical panel ($R/h = 20$, $R/b = 2$, $b/a = 1$, type D meshes).

Mesh	NEQ	Ω_6	Ω_7	Ω_8	Ω_9	Ω_{10}
8×8D	678	0.770373	0.770373	0.805719	0.911450	0.911450
12×12D	1590	0.744312	0.744312	0.794981	0.845463	0.845463
16×16D	2886	0.736167	0.736167	0.790634	0.826941	0.826941
20×20D	4566	0.732604	0.732604	0.788544	0.819276	0.819276
24×24D	6630	0.730751	0.730751	0.787414	0.815440	0.815440
[5.3][1]		0.73299	0.73299	0.77902	N/A	N/A
[5.4][2]		0.72609	0.72609	0.77493	N/A	N/A

[1] Thin shell results.
[2] Thick shell results.

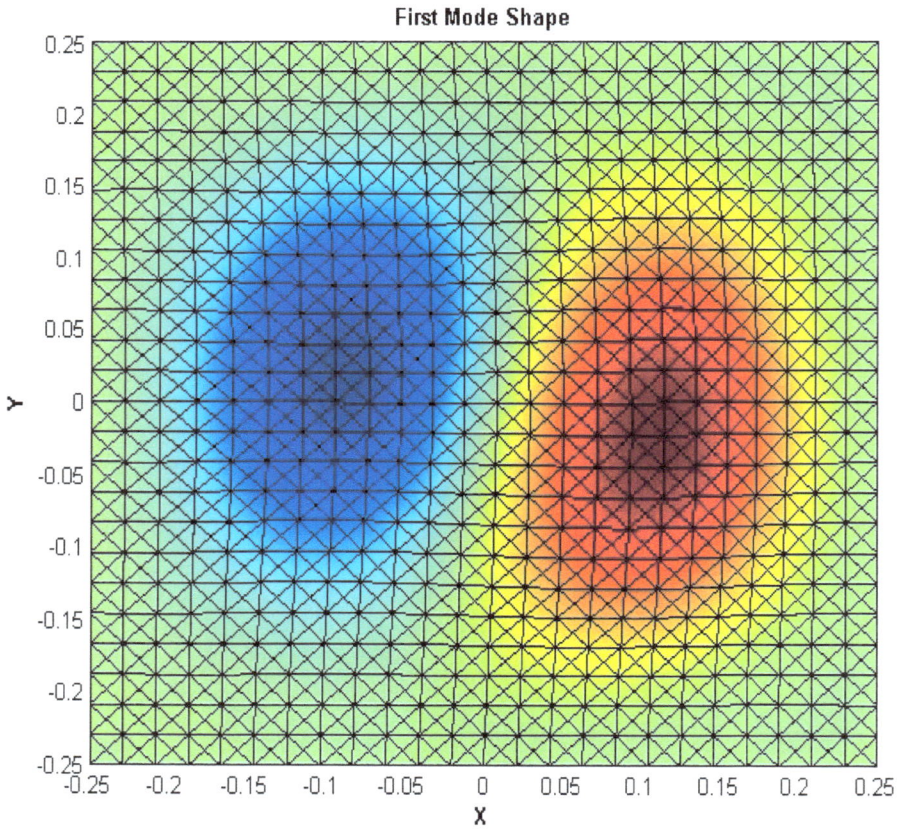

Figure 5.7(a): First mode shape of fully clamped spherical panel.

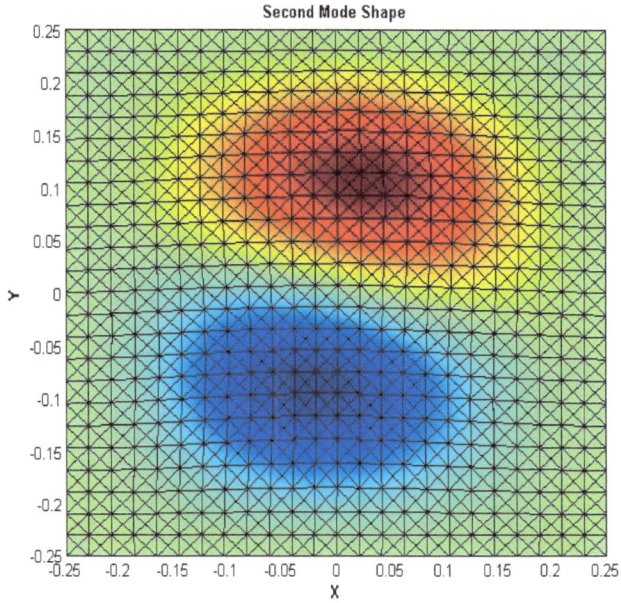

Figure 5.7(b): Second mode shape of fully clamped spherical panel.

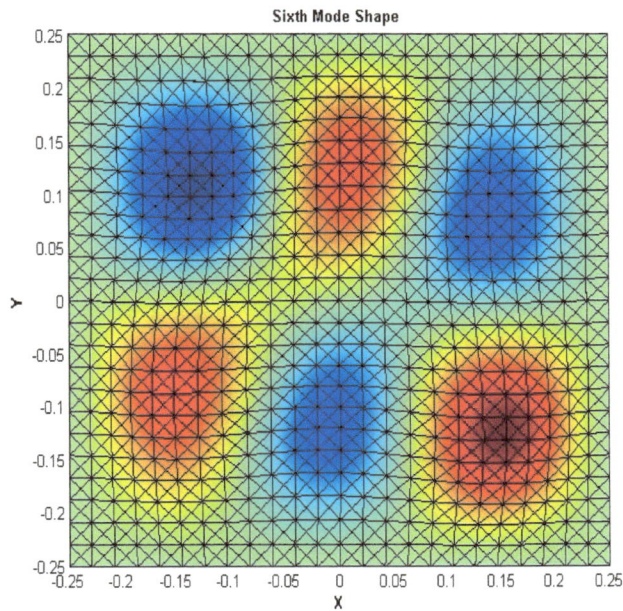

Figure 5.7(c): Sixth mode shape of fully clamped spherical panel.

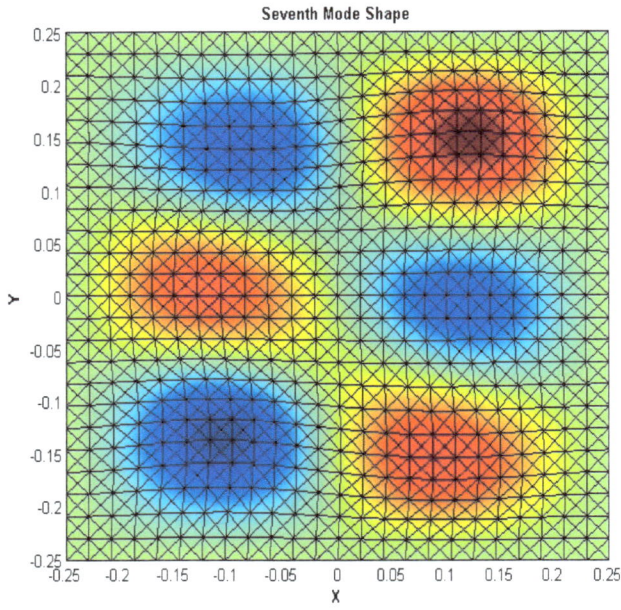

Figure 5.7(d): Seventh mode shape of fully clamped spherical panel.

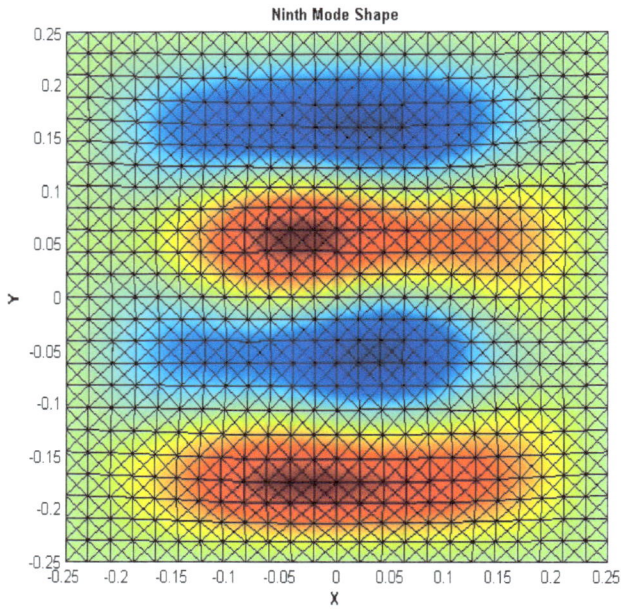

Figure 5.7(e): Ninth mode shape of fully clamped spherical panel.

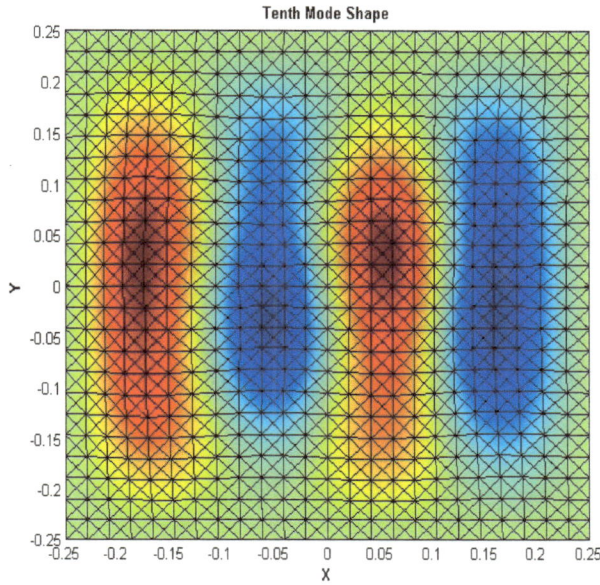

Figure 5.7(f): Tenth mode shape of fully clamped spherical panel.

5.3. HEMISPHERICAL PANEL

The vibration of a hemispherical panel shown in Fig. **5.8** is investigated in this section. The panel has an azimuthal angle of 120° and a polar angle of 30°. It is either clamped or simply supported at edges *BC* and *CD* in Fig. **5.8**. When simply supported, edges *BC* and *CD* are imposed the following boundary conditions of *U* = *V* = *W* = 0. On the other hand, if *BC* and *CD* are clamped, all their DOF are constrained. The dimensions and material properties are: *R* = 1.0 m, *h* = 0.1 m, *E* = 200 GPa, *v* = 0.3 and ρ_0 = 7830 kg/m³. The computed results are presented in Tables **5.8** for the clamped case and **9** for the simply supported panel. The frequency parameter is $\Omega = \omega\sqrt{\rho_0(1 - v^2)/E}$. In the tables, mesh designation is that the first integer means the number of divisions in the longitudinal direction while the second integer shows the number of divisions in the latitudinal direction. The tables also include results of Ref. [5.5]. These results were obtained by the generalized differential quadrature (GDQ) method. However, the boundness of the numerical solutions was not established in [5.5]. The following discussions may be appropriate.

(i)   Since $R/h = 10$, it is very likely that the present results are lower bound solutions (see Sec. 3.2 and Sec. 4.5, for example).

(ii) Comparing results from the finest mesh used, 27×20D, with those of [5.5], the present results have lower values over all ten natural frequencies. It is seen from Table **5.8**, that the first mode has a discrepancy of -2.443%, and the eighth mode has a difference of -6.290%.

(iii) Similarly, for the simply supported case the differences in the first and second modes are, -2.039% and -2.146%, respectively.

(iv) The first five mode shapes of the clamped panel are plotted in Fig. **5.9**.

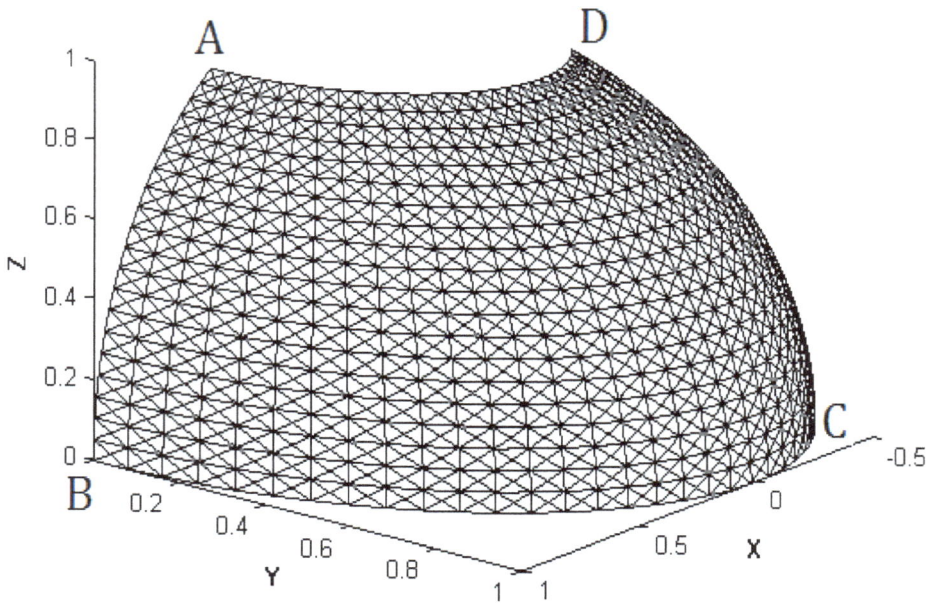

Figure 5.8: Spherical panel clamped or simply supported at BC and CD.

Table 5.8: First ten frequency parameters $\Omega = \omega\sqrt{\dfrac{\rho_0\left(1-\nu^2\right)}{E}}$ of clamped hemispherical panel.

Mesh	NEQ	Ω_1	Ω_2	Ω_3	Ω_4	Ω_5
7×5D	420	0.379974	0.526325	0.820583	1.02846	1.21137
13×10D	1650	0.371813	0.512987	0801380	1.00280	1.17737
19×14D	3192	0.369853	0.511048	0.797651	0.996600	1.17168
24×18D	5184	0.369036	0.509707	0.795662	0.994695	1.16865
27×20D	6480	0.368728	0.509510	0.795166	0.993757	1.16806
[5.5]		0.377734	0.529971	0.815963	1.02231	1.21010
Mesh	NEQ	Ω_6	Ω_7	Ω_8	Ω_9	Ω_{10}
7×5D	420	1.48472	1.48638	1.58373	1.66573	1.85062
13×10D	1650	1.41486	1.43667	1.51584	1.62113	1.77752
19×14D	3192	1.40799	1.42572	1.50438	1.61452	1.76532
24×18D	5184	1.40151	1.42239	1.49906	1.61015	1.75908
27×20D	6480	1.40126	1.42091	1.49770	1.60956	1.75764
[5.5]		1.47003	1.50703	1.59823	1.66606	1.84584

Table 5.9: First five frequency parameters $\Omega = \omega\sqrt{\dfrac{\rho_0\left(1-\nu^2\right)}{E}}$ of simply supported hemispherical panel.

Mesh	NEQ	Ω_1	Ω_2	Ω_3	Ω_4	Ω_5
7×5D	451	0.346373	0.456190	0.762933	0.942775	1.11404
13×10D	1618	0.338750	0.450430	0.750468	0.930060	1.09686
19×14D	3174	0.337997	0.448591	0.747990	0.927553	1.09372
24×18D	5288	0.337711	0.447672	0.747125	0.926920	1.09213
27×20D	6596	0.337714	0.447376	0.746881	0.926600	1.09173
[5.5]		0.344743	0.457185	0.759268	0.945132	1.11143

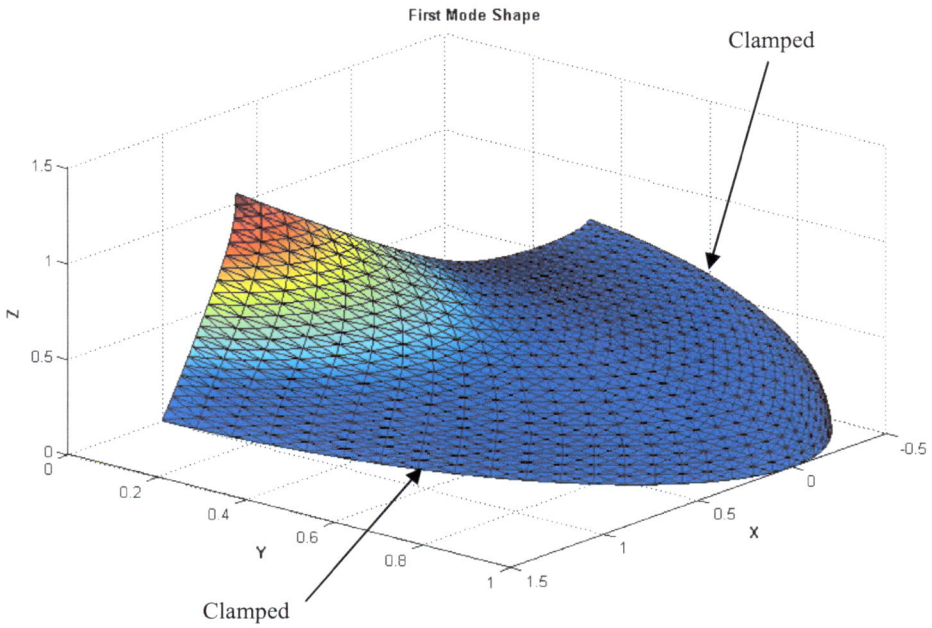

Figure 5.9(a): First mode shape of clamped hemispherical panel.

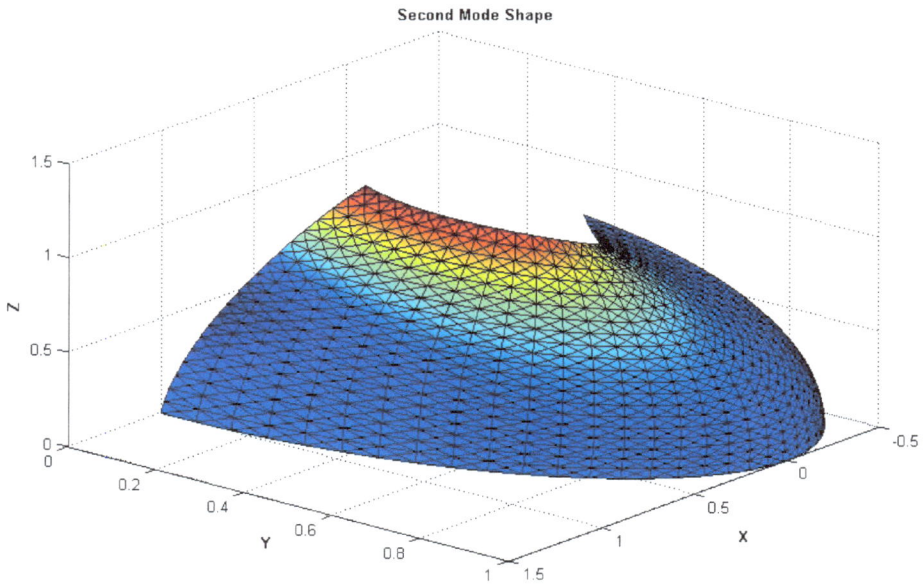

Figure 5.9(b): Second mode shape of clamped hemispherical panel.

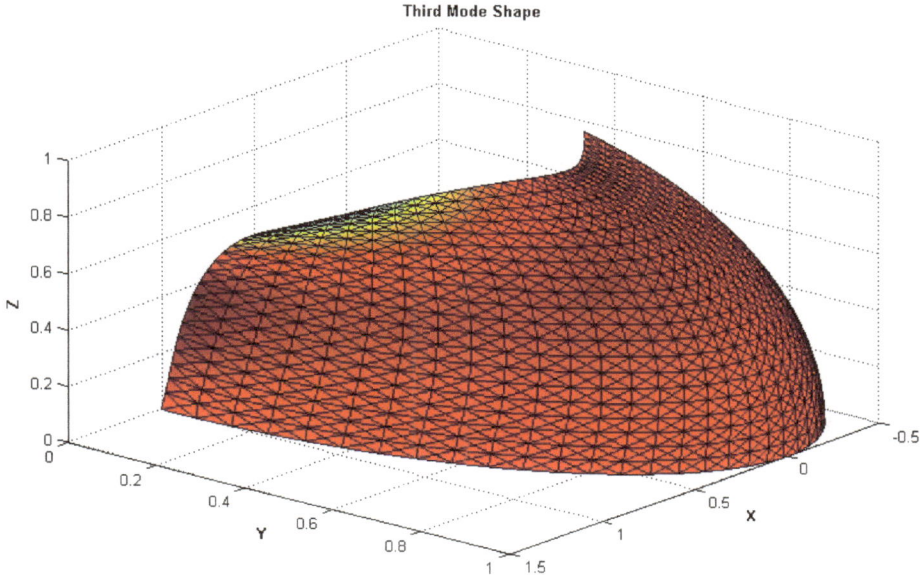

Figure 5.9(c): Third mode shape of clamped hemispherical panel.

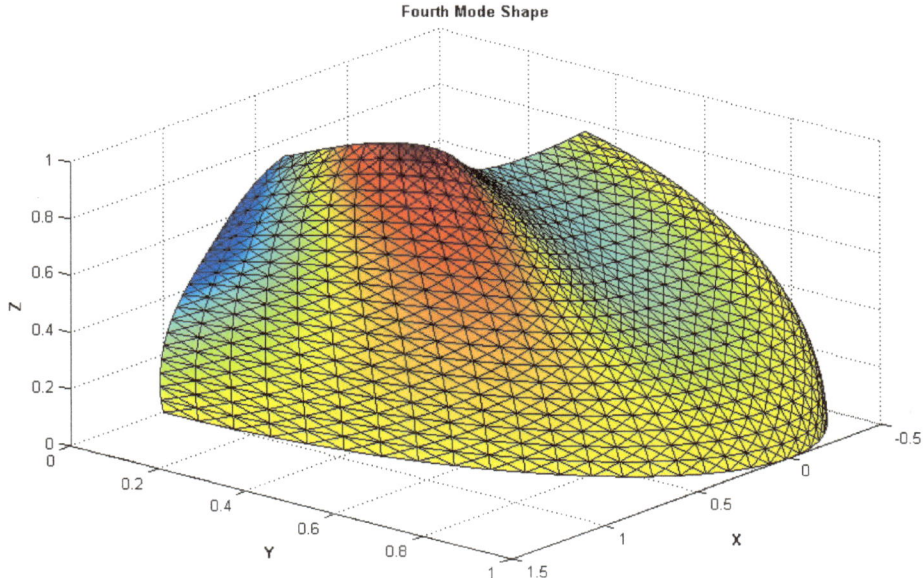

Figure 5.9(d): Fourth mode shape of clamped hemispherical panel.

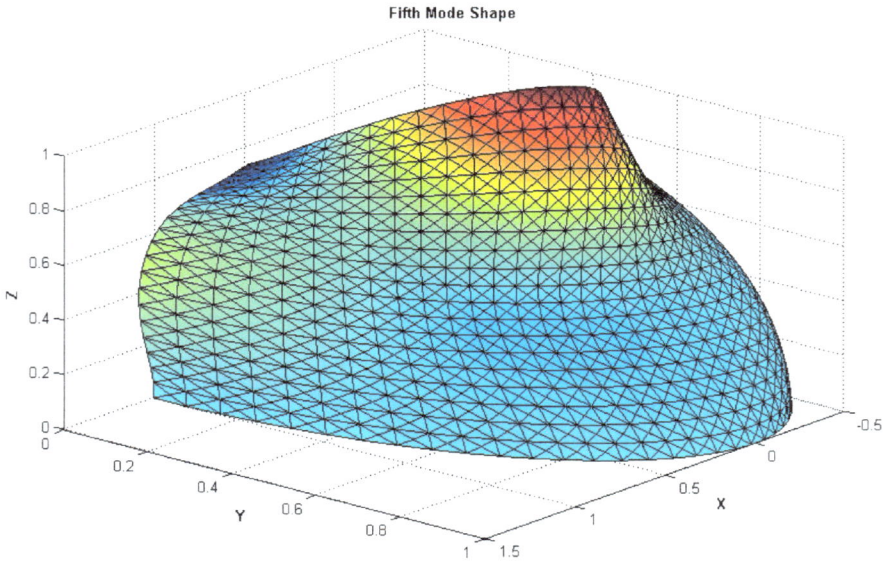

Figure 5.9(e): Fifth mode shape of clamped hemispherical panel.

5.4. CLAMPED HEMISPHERICAL SHELL

Now the vibration characteristics of a hemispherical shell are investigated. This deep shell, shown in Fig. **5.10**, is clamped at the base. The following dimensions and material properties are used in the finite element computations: a polar angle = 30°, R = 1.0 m, h = 0.1 m, E = 200 GPa, v = 0.3 and ρ_0 = 7830 kg/m³. The entire shell is discretized. For the nodes at the base, all their nodal DOF are constrained.

Computed frequency parameters are listed in Table **5.10** where $\Omega = \omega\sqrt{\rho_0(1-v^2)/E}$ and mesh designations are similar to those of the hemispherical panel in the last section. Remarks similar to those in the last section can be made, and are not repeated here for brevity. Between the 48×12D results and those from [5.5], the largest discrepancy is -2.731% at the tenth natural frequency. Finally, the first, third, fifth, seventh, eighth and tenth mode shapes are given in Fig. **5.11**. The second, fourth, sixth and ninth mode shapes are similar, respectively, to the first, third, fifth and eighth modes by rotating 90° clockwise about the Z-axis and therefore they are not included here for brevity. It is also noted that the seventh mode is an axi-symmetric mode.

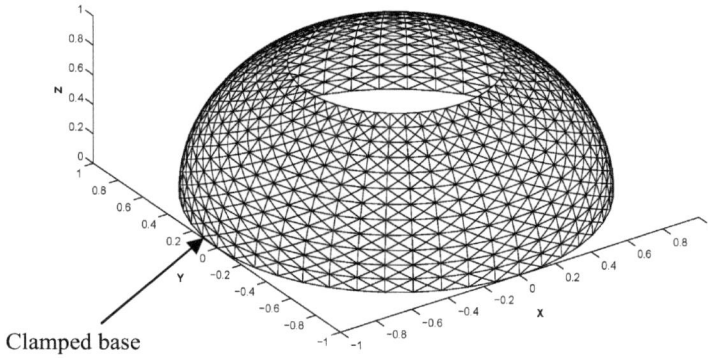

Clamped base

Figure 5.10: Hemispherical shell clamped at base with 48×12D mesh shown.

Table 5.10: First ten frequency parameters $\Omega = \omega\sqrt{\dfrac{\rho_0\left(1-\nu^2\right)}{E}}$ of clamped hemispherical shell.

Mesh	NEQ	Ω_1	Ω_2	Ω_3	Ω_4	Ω_5
20×5D	1200	0.427070	0.427070	0.671992	0.671992	0.733610
32×8D	3072	0.420360	0.420360	0.669108	0.669108	0.714332
40×10D	4800	0.418699	0.418699	0.668360	0.668360	0.709750
48×12D	6912	0.417730	0.417730	0.667913	0.667913	0.707168
[5.5]		0.426643	0.426643	0.672193	0.672193	0.726092
Mesh	**NEQ**	Ω_6	Ω_7	Ω_8	Ω_9	Ω_{10}
10×5D	1200	0.733610	0.935889	1.03175	1.03175	1.12488
20×10D	3072	0.714332	0.924837	1.01607	1.01607	1.09613
28×14D	4800	0.709750	0.922427	1.01244	1.01244	1.08954
36×18D	6912	0.707168	0.921141	1.01042	1.01042	1.08593
[5.5]		0.726092	0.936907	1.02690	1.02690	1.11559

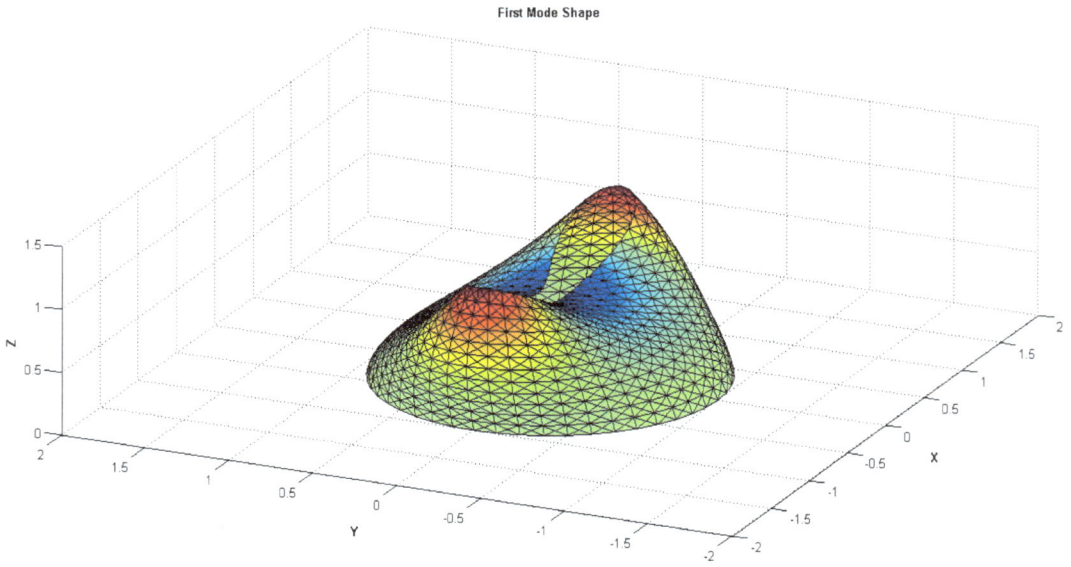

Figure 5.11(a): First mode shape of hemispherical shell [$\Omega = 0.417730$].

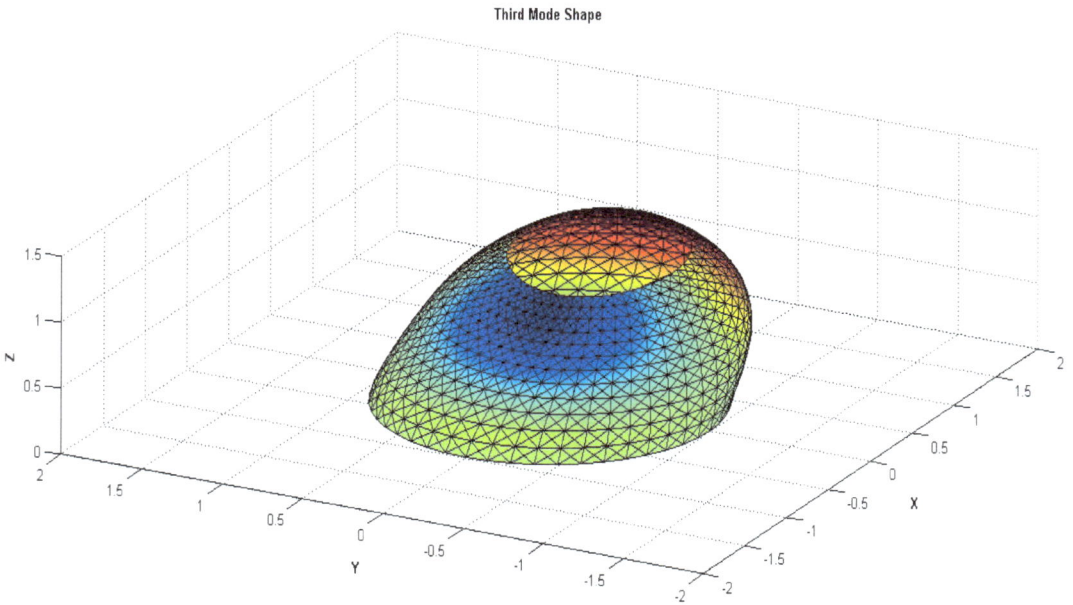

Figure 5.11(b): Third mode shape of hemispherical shell [$\Omega = 0.667913$].

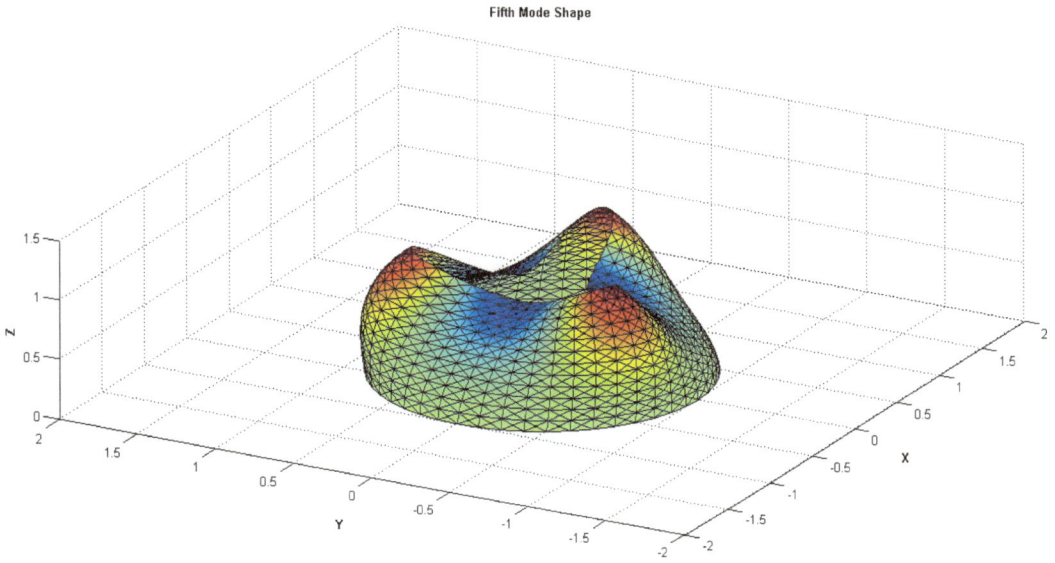

Figure 5.11(c): Fifth mode shape of hemispherical shell [$\Omega = 0.707168$].

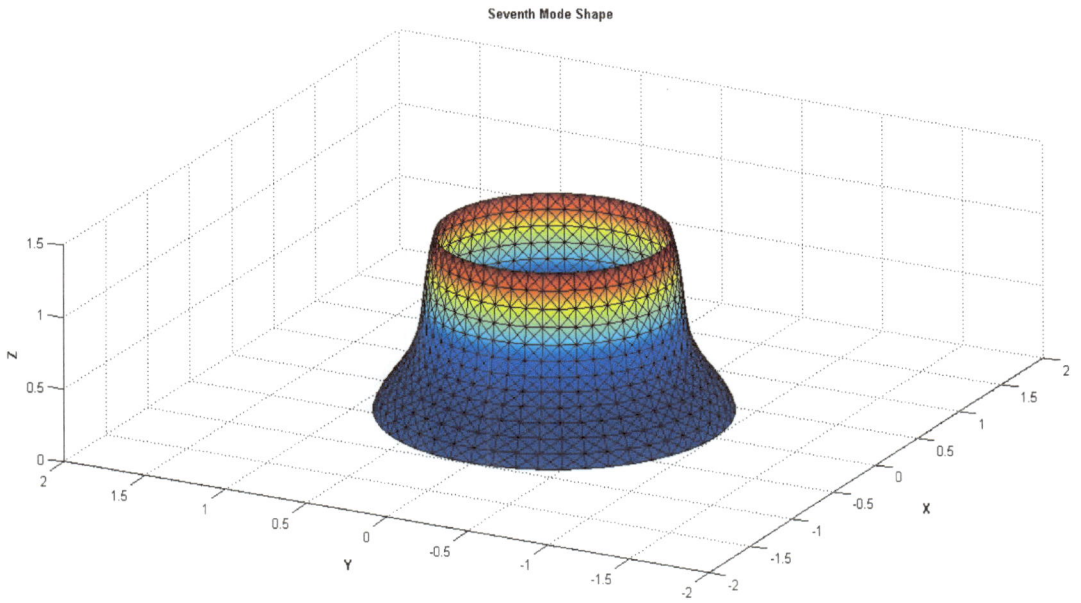

Figure 5.11(d): Seventh mode shape of hemispherical shell [$\Omega = 0.921141$].

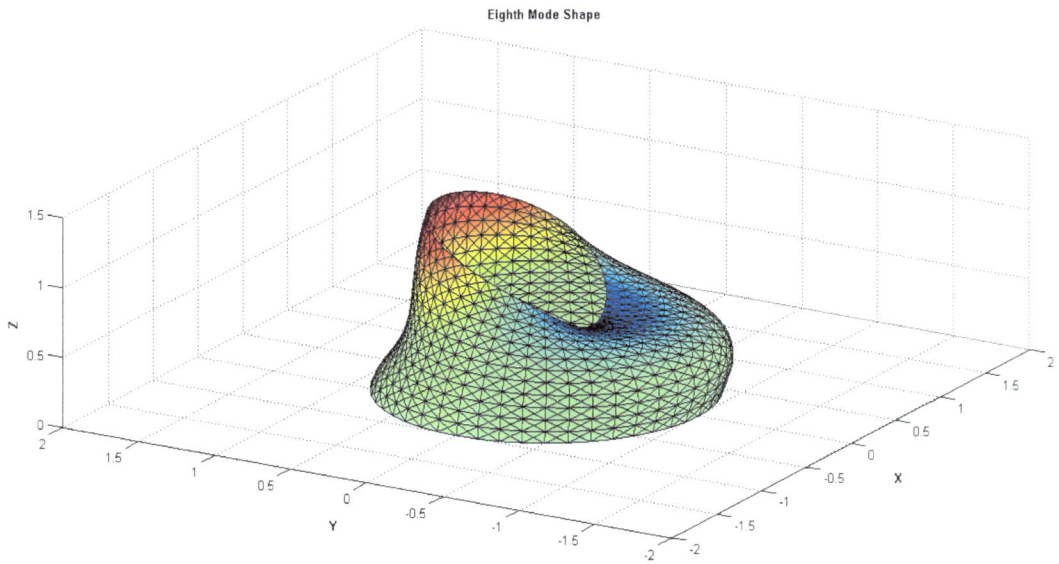

Figure 5.11(e): Eighth mode shape of hemispherical shell [$\Omega = 1.01042$].

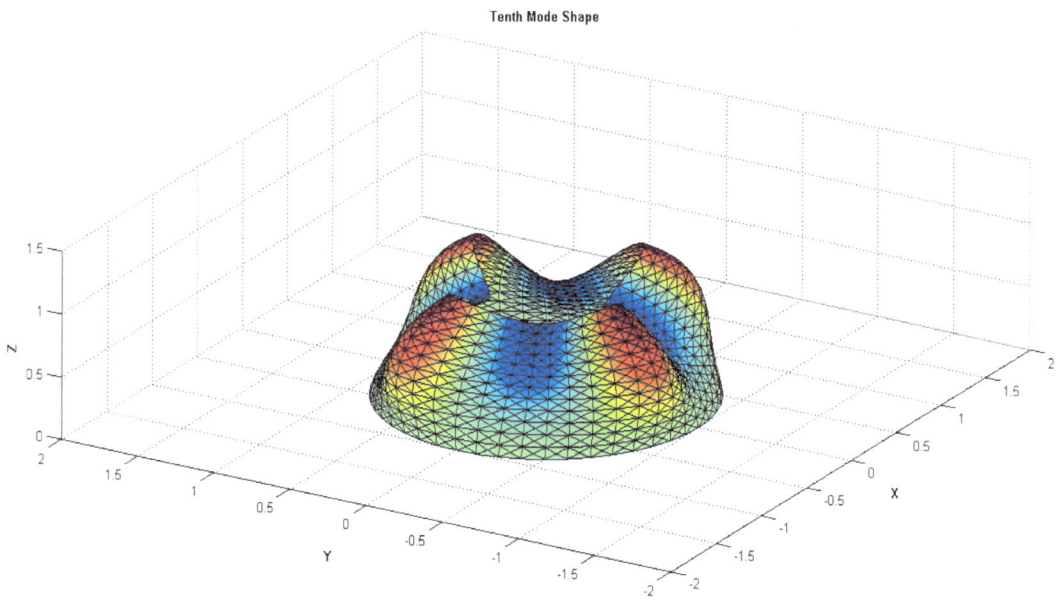

Figure 5.11(f): Tenth mode shape of hemispherical shell [$\Omega = 1.08593$].

REFERENCES

[5.1]   K.M. Liew, and C.W. Lim, "A higher-order theory for vibration analysis of curvilinear thick shallow shells with constrained boundaries", *J. Vibr. Control*, vol. 1, pp. 15-39, January 1995.

[5.2]   K.S.S. Ram, and T.S. Babu, "Free vibration of composite spherical shell cap with and without a cutout", *Comp. Struct.*, vol. 80, pp. 1749-1756, September 2002.

[5.3]   K.M. Liew, and C.W. Lim, "Vibratory characteristics of cantilevered rectangular shallow shells of variable thickness", *AIAA J.*, vol. 32, pp. 387-396, February 1994.

[5.4]   K.M. Liew, L.X. Peng, and T.Y. Ng, "Three-dimensional vibration analysis of spherical shell panels subjected to different boundary conditions", *Int. J. Mech. Sci.*, vol. 44, pp. 2103-2117, October 2002.

[5.5]   F. Tornabene, and E. Viola, "Vibration analysis of spherical structural elements using the GDQ method", *Int. J. Comp. Math. Applications*, vol. 53, pp. 1538-1560, May 2007.

Send Orders for Reprints to reprints@benthamscience.net

CHAPTER 6

Vibration Analysis of Box Structures

Abstract: Vibration characteristics of box structures, generally known as rectangular prismatic shell structures, are studied in this chapter. To limit the scope in this book, only single-cell box structure and double-cell box structures are presented. In these two types of shell structures the first ten frequency parameters or natural frequencies along with their corresponding mode shapes are included.

Keywords: Vibration analysis, box-structures, single-cell, double-cell.

6.1. INTRODUCTION

In Chapters 2 through 5, vibration characteristics of plates, shells with single curvature, and shells with double curvatures have been presented. A category of special shell structures is considered in this chapter. This category of structures can be regarded as shell structures with double zero curvatures and is generally known as rectangular prismatic shells. To limit the scope of the present investigation, only single and double cell box structures are included in the following sections.

6.2. SINGLE-CELL BOX STRUCTURE

As mentioned in Sec. 2.1, the advantage of having DDOF in the nodal DOF of a shell finite element is able to avoid singular global stiffness matrix without invoking complex schemes or artificial parameters. In this section, a single-cell box structure (SCBS) is selected to demonstrate the applicability of one of the shell elements to structures whose mid-surface contains sharp junctures. As shown in Fig. **6.1(a)**, the single-cell box structure is clamped at the base. The length is $L = 1.0$ m. Dimensions B_1 and B_2 are $B_1/L = 0.33$ and $B_2/L = 0.25$. The thickness is uniform and given by $h/L = 0.005$. Material properties are: $E = 200$ GPa, $v = 0.32$ and $\rho_0 = 7830$ kg/m^3.

To compare with the symmetric (about the mid-vertical plane) modes reported in Refs. [6.1, 6.2], half of the SCBS is first discretized. Results are summarized in Table **6.1**. The three integers in mesh designators represent, respectively, the numbers of divisions in the horizontal ($2B_1$), vertical ($2B_2$) and length (L) directions. It is seen that the finest mesh 8×12×24D gives higher natural frequencies than those reported in [6.1, 6.2] with discrepancy ranging from 5.760% to 6.688%.

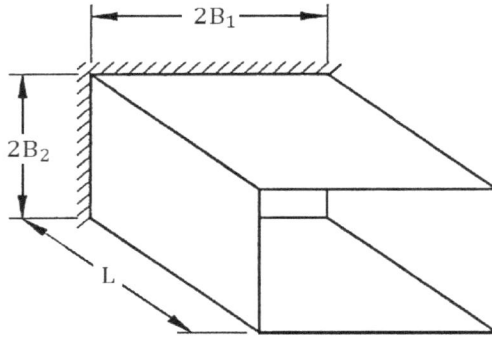

Figure 6.1(a): Clamped single-cell box structure (SCBS).

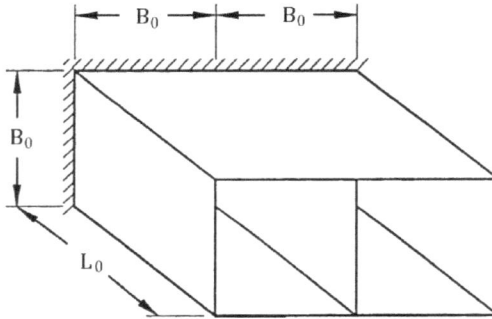

Figure 6.1(b): Clamped double-cell box structure (DCBS).

Having applied the symmetry property to the analysis above, now the entire box is modeled and computed results are tabulated in Table **6.2**. The first ten mode shapes are shown in Fig. **6.2**. With reference to the presented results it is found that,

(i) the full box representation is more accurate than the half-box one, given the same or similar NEQ; for example, the first natural frequency has a discrepancy of only 0.07505%, at a NEQ of 6912, instead of 8064; and

(ii) the fifth and seventh modes are symmetric modes about the mid-horizontal plane and cannot be readily compared with results from [6.1-6.2] since the latter contain symmetric modes about the mid-vertical plane only.

Table 6.1: First ten frequency parameters $\Omega = \omega L\sqrt{\dfrac{\rho_0\left(1-\nu^2\right)}{E}}$ of SCBS (half model).

Mesh	NEQ	Ω_1	Ω_2	Ω_3	Ω_4	Ω_5
3×5×10D	1320	0.0468032	0.0623732	0.0786559	0.0893336	0.120216
4×6×12D	2016	0.0463984	0.0612394	0.0779284	0.0877844	0.116998
5×7×15D	3060	0.0461891	0.0606861	0.0775078	0.0870136	0.115313
6×10×20D	5280	0.0460390	0.0602866	0.0771845	0.0864135	0.113567
8×12×24D	8064	0.0459378	0.0600122	0.0770024	0.0860458	0.112822
[6.1]		0.0434	0.0565	0.0728	0.0811	0.106
[6.2]		0.0437	0.0569	0.0736	0.0816	0.106
Mesh	**NEQ**	Ω_6	Ω_7	Ω_8	Ω_9	Ω_{10}
3×5×10D	1320	0.140787	0.144916	0.147495	0.199494	0.237813
4×6×12D	2016	0.139190	0.140889	0.145190	0.193996	0.233865
5×7×15D	3060	0.137871	0.138997	0.143700	0.191073	0.230473
6×10×20D	5280	0.136405	0.137451	0.142481	0.188299	0.227834
8×12×24D	8064	0.135554	0.136989	0.141917	0.187029	0.226891
[6.1]		0.127	0.129	0.134	0.175	0.213
[6.2]		0.128	0.130	0.135	0.177	N/A

Table 6.2: First ten frequency parameters $\Omega = \omega L\sqrt{\dfrac{\rho_0\left(1-\nu^2\right)}{E}}$ of SCBS (full model).

Mesh	NEQ	Ω_1	Ω_2	Ω_3	Ω_4	Ω_5
6×5×10D	2640	0.0443409	0.0590918	0.0745178	0.0846338	0.0863505
8×6×12D	4032	0.0439574	0.0580176	0.0738287	0.0832661	0.0851664
10×8×16D	6912	0.0437328	0.0574223	0.0733572	0.0823028	0.0841665
[6.1]		0.0434	0.0565	0.0728	0.0811	N/A
[6.2]		0.0437	0.0569	0.0736	0.0816	N/A
Mesh	**NEQ**	Ω_6	Ω_7	Ω_8	Ω_9	Ω_{10}
6×5×10D	2640	0.113891	0.115308	0.133380	0.137292	0.139735
8×6×12D	4032	0.110843	0.113517	0.131867	0.133477	0.137551
10×8×16D	6912	0.108643	0.112141	0.130254	0.131090	0.135808
[6.1]		0.106	N/A	0.127	0.129	0.134
[6.2]		0.106	N/A	0.128	0.130	0.135

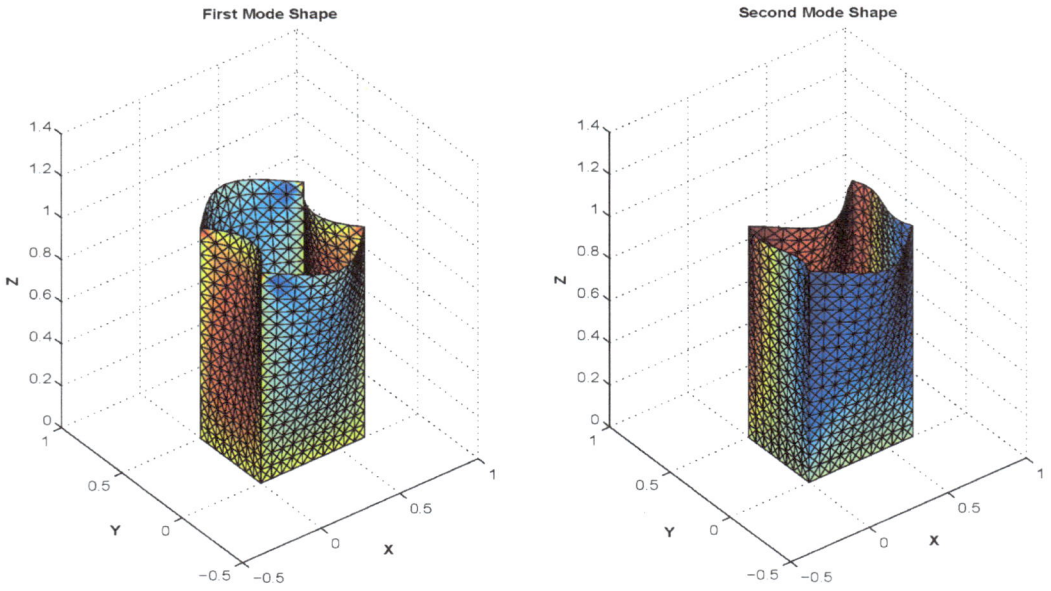

Figure 6.2(a): First and second mode shapes of SCBS.

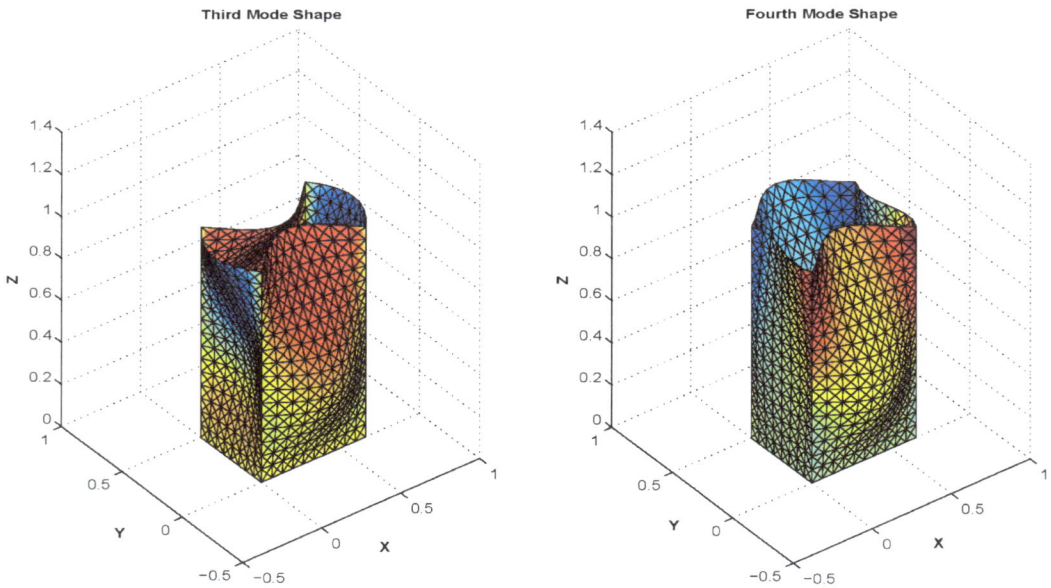

Figure 6.2(b): Third and fourth mode shapes of SCBS.

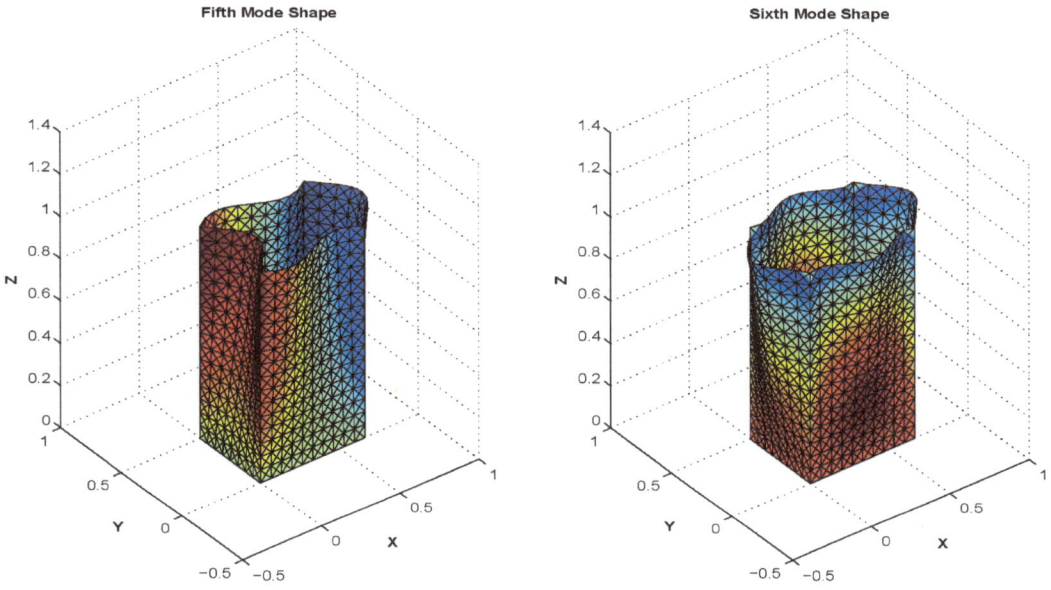

Figure 6.2(c): Fifth and sixth mode shapes of SCBS.

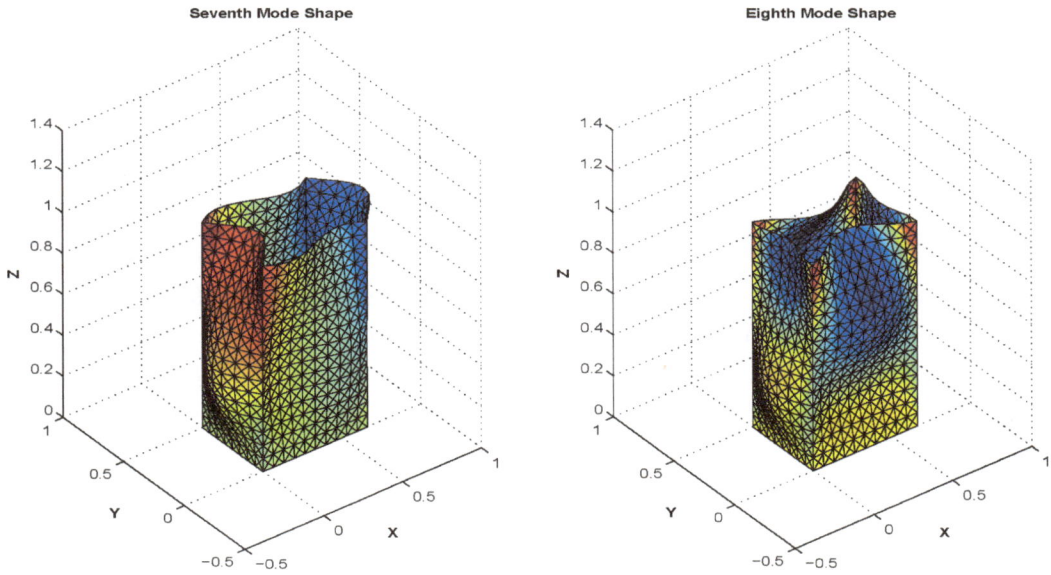

Figure 6.2(d): Seventh and eighth mode shapes of SCBS.

Ninth Mode Shape   Tenth Mode Shape

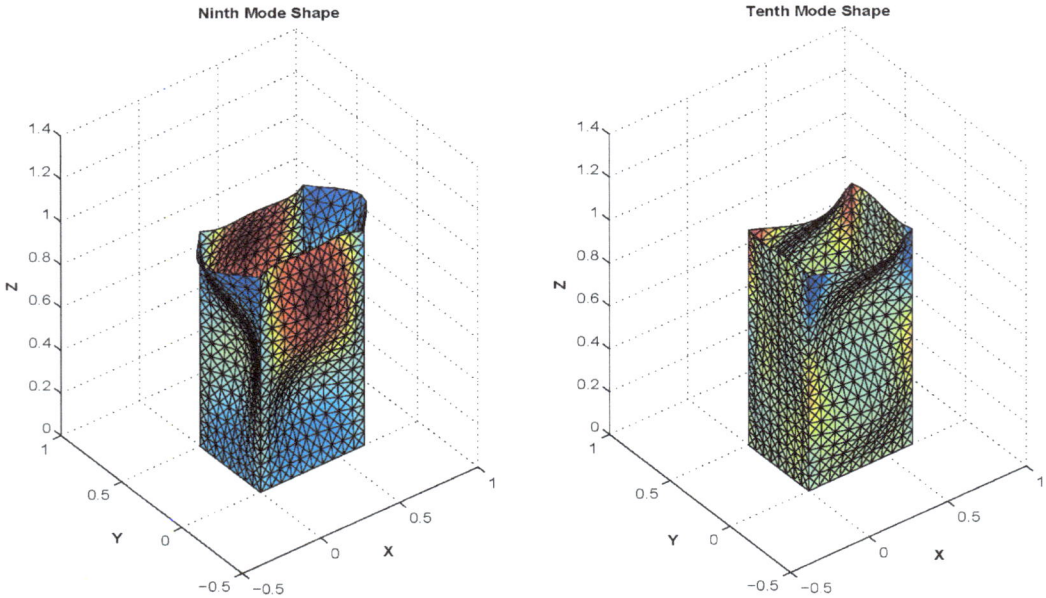

Figure 6.2(e): Ninth and tenth mode shapes of SCBS.

6.3. DOUBLE-CELL BOX STRUCTURE

In this section, the double-cell box structure (DCBS) shown in Fig. **6.1(b)** is studied. This is a logical extension to the SCBS. The DCBS has the following material properties: E = 207 GPa (30×10^6 psi), v = 0.3 and ρ_0 = 7850 kg/m^3 (0.000735 lb·sec^2/in^4). Its thickness is uniform at h = 12.7 mm (0.5 in). Other geometrical properties are: B_o = 1.270 m (50 in), and L_o = 2.540 m (100 in). The first ten computed natural frequencies are given in Table **6.3**, and their corresponding mode shapes in Fig. **6.3**. Compared with results reported in [6.1], the 20×10×20B mesh gives a discrepancy of 1.422% for the fundamental natural frequency. The highest discrepancy is 6.523% at the ninth natural frequency, while the first five natural frequencies are accurately predicted with less than 5% discrepancy.

In passing, it may be noted that the referenced results from [6.1] were obtained by using 56 spline elements, every one of which has 63 DOF.

Table 6.3: First ten natural frequencies (in rad/s) of DCBS.

Mesh	NEQ	ω_1	ω_2	ω_3	ω_4	ω_5
10×5×10B	2040	146.185	184.697	209.732	237.960	265.152
12×6×12B	2952	137.411	173.783	197.324	223.340	248.875
14×7×14B	4032	133.336	168.551	191.325	216.242	224.882
16×8×16B	5280	131.200	165.717	188.037	212.298	212.365
20×10×20B	8280	129.232	162.985	184.803	201.183	208.455
[6.1]		127.42	159.81	180.88	192.42	203.20
Mesh	**NEQ**	ω_6	ω_7	ω_8	ω_9	ω_{10}
10×5×10B	2040	298.150	331.518	333.694	334.608	358.635
12×6×12B	2952	248.957	282.399	305.254	309.531	311.506
14×7×14B	4032	241.038	256.987	279.637	297.769	298.200
16×8×16B	5280	236.633	242.894	264.459	288.181	291.270
20×10×20B	8280	229.597	232.196	249.661	271.356	284.719
[6.1]		217.76	225.73	235.61	254.74	274.42

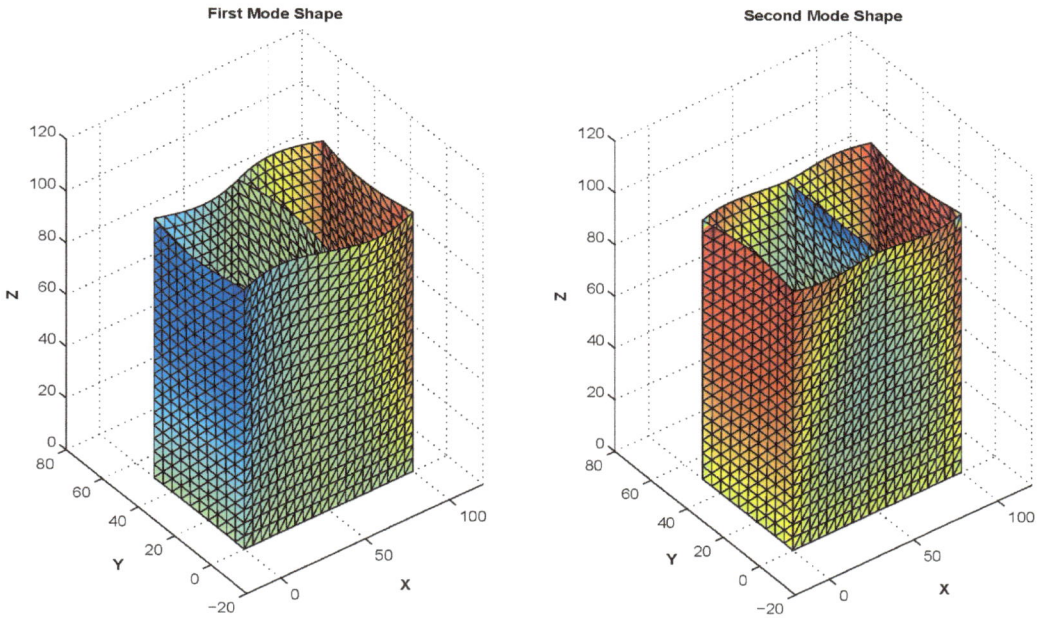

Figure 6.3(a): First and second mode shapes of DCBS.

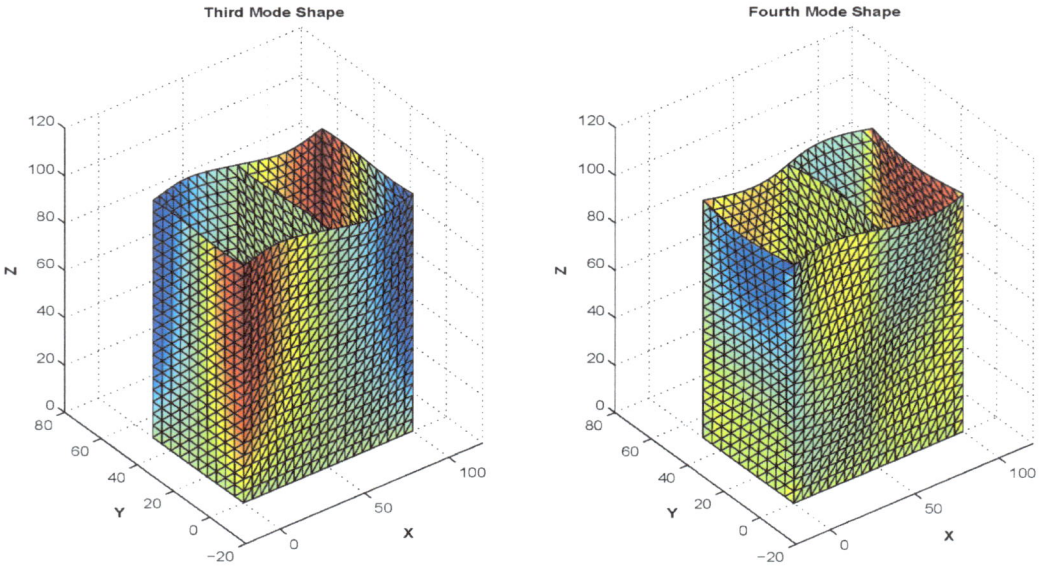

Figure 6.3(b): Third and fourth mode shapes of DCBS.

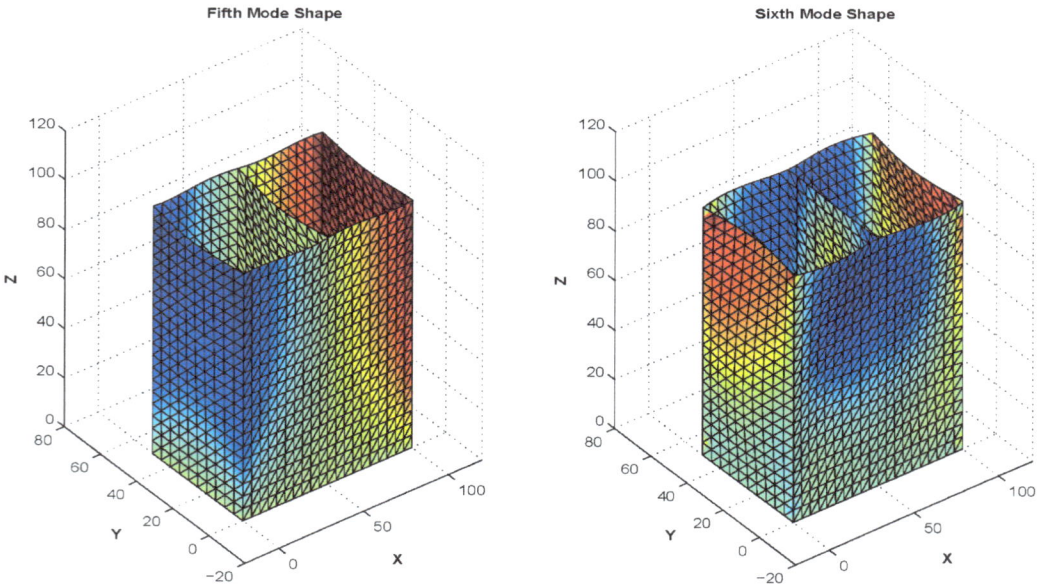

Figure 6.3(c): Fifth and sixth mode shapes of DCBS.

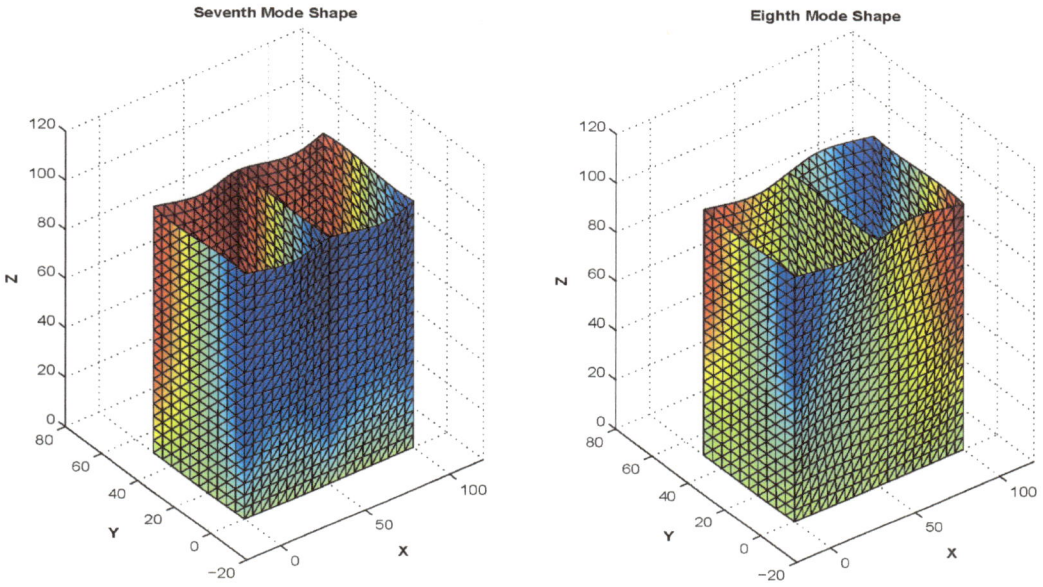

Figure 6.3(d): Seventh and eighth mode shapes of DCBS.

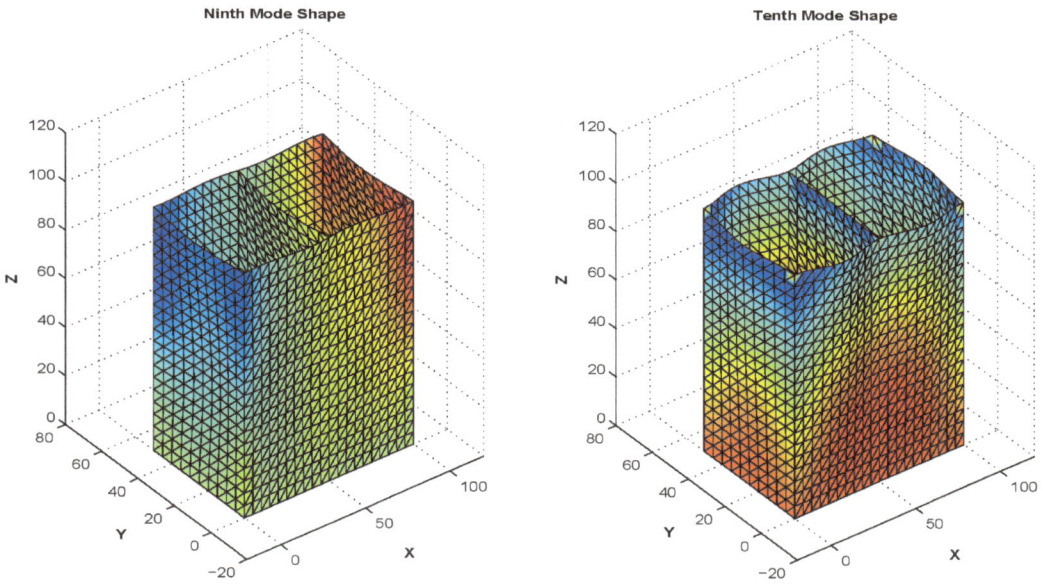

Figure 6.3(e): Ninth and tenth mode shapes of DCBS.

6.4. CLOSURE

In Sec. 6.2 and 6.3, the SCBS and DCBS have been respectively studied. It is of great interest to the structural dynamicists to have knowledge of vibration characteristics of more multi-cell box structures (MCBS's). However, the number of DOF and computational time increase rapidly with growing number of cells. Besides, it is beyond the scope of this book and therefore, it is not pursued further in this chapter.

REFERENCES

[6.1]   S.C. Fan, and M.H. Luah, "Free vibration analysis of arbitrary thin shell structures by using spline finite element", *J. Sound Vibr.*, vol. 179, pp. 763-776, February 1995.

[6.2]   T. Irie, G. Yamada, and Y. Kobayashi, "Free vibration of an oblique rectangular prismatic shell", *J. Sound Vibr.*, vol. 102, pp. 501-513, October 1985.

Send Orders for Reprints to reprints@benthamscience.net
Vibration and Nonlinear Dynamics of Plates and Shells, 2014, 111-148   **111**

Mixed Formulation Based Three-Node Flat Triangular Shell Elements for Nonlinear Dynamics

Abstract: To further theobjective of investigating the vibration characteristics and dynamic responses of complicated shell structures with geometrical and material nonlinearities, this chapter presents the development of mixed formulation based three-node flat triangular shell elements suitable for the general nonlinear analysis of thin to moderately thick shells. Section 7.1 gives a brief outline of the features of the shell elements. Section 7.2 presents the incremental variational principle and its linearization. The derivations of the consistent element stiffness matrices and the consistent element mass matrices are dealt with in Sections 7.3 and 7.4. The constitutive relations of elastic and elasto-plastic materials with small as well as finite strain deformations are given in Section 7.5. The last two sections, Sections 7.6 and 7.7, respectively, are concerned with configuration and stress updating, and numerical algorithms.

Keywords: Mixed formulation, three-node shell elements, dynamics, nonlinear, elastic and elasto-plastic.

7.1. INTRODUCTION AND OVERVIEW

This chapter presents the mixed formulation based three-node triangular flat shell elements in the context of general nonlinear dynamic analysis of structures. Some of the important features of the elements are as follows.

(i)   The updated Lagrangian formulation and the incremental Hellinger-Reissner variational principle are employed. The independently assumed fields include the incremental displacements and incremental strains. Accordingly, the incremental second Piola-Kirchhoff stress and the incremental Washizu strain are selected as the incremental stress and strain measures.

(ii) Two versions of the nonlinear element stiffness matrices are developed. These are the director version and the simplified version. In the director version, it is assumed that for every node on the shell mid-surface the director can be uniquely defined. The stiffness matrices are found to be dependent of the current position of the

Meilan Liu and Cho W. S. To

director. Thus, it requires the updating of the director at every time step. The simplified version, on the other hand, is useful for cases where the director is not unique, or is difficult to determine. For brevity, only the derivation of director version of the matrices is presented. The simplified version can be easily deducted from the director version.

(iii) To be consistent with element stiffness matrices, the element consistent mass matrices have their director version and simplified version, depending on whether the director is uniquely defined. The consistent mass matrix is defined with respect to the reference configuration. Therefore, it is to be calculated at every time step, since mass density, thickness of the shell, positions of mid-surface nodes, and directors for the director version of formulation, change as the shell deforms.

(iv) Material nonlinearity is of the elasto-plastic type with isotropic strain hardening. The J_2 flow theory of plasticity, in conjunction with Ilyushin's yield criterion, is employed. The non-layered approach is adopted, in order to simplify the derivation of stiffness matrix and to facilitate the acquisition of explicit expressions for the stiffness matrices.

(v) For homogeneous, isotropic and linearly elastic materials, as well as elasto-plastic materials with isotropic strain hardening, formulations for small and large strain applications are included.

(vi) Emphasis is also placed on the updating of director field, director velocity field and director acceleration field. Schemes to update stress and to transform the second Piola-Kirchhoff stress to the Cauchy stress are considered.

(vii)Explicit expressions for the element stiffness matrices (linear and nonlinear), element consistent mass matrix and element pseudo-force vector are obtained. These expressions are written in terms of geometry and state of stress of the reference configuration which would be available from solutions of the previous time step.

(viii) The two elements that were presented in Chapter 2, $AT+(k_t^1)$' and $AT+(k_t^3)$', can be viewed as the simplified version of the element linear stiffness matrix k_L of Sub-sec. 7.3.4.

7.2. INCREMENTAL   VARIATIONAL   PRINCIPLE   AND   ITS LINEARIZATION

As in any nonlinear analysis, the fundamental difficulty is that the configuration of a body at time "$t+\Delta t$" is unknown. In seeking for approximate solution of this nonlinear analysis, the incremental formulation assumes that the static and kinematic variables in the current configuration C^t are known. Their values in an unknown neighbouring configuration $C^{t+\Delta t}$ at a later time "$t+\Delta t$" are then determined from these known solutions. The starting point of such an incremental analysis is an appropriate incremental variational principle, which is, in the present study, the Hellinger-Reissner variational principle in the incremental form. It has two independently assumed fields, the incremental displacements (and rotations) and the incremental strains. The principle takes the following form [7.1, 7.2]:

$$\Delta\pi_{HR}\left(\Delta u,\Delta e\right)=$$
$$\int_{V^t}\left[-\tfrac{1}{2}\left(\Delta e\right)^T\tilde{D}\left(\Delta e\right)+\sigma^T\left(\Delta\overline{e}\right)+\left(\Delta e\right)^T\left(\Delta e\right)^T\tilde{D}\left(\Delta\overline{e}\right)\right]dV-W^{t+\Delta t}\cdots \tag{7.1}$$

where the integration is carried out over the reference volume V^t, and

Δu is the vector of assumed incremental displacement;

Δe is the vector of assumed incremental Green strain;

$\Delta\overline{e}$ is the vector of incremental Washizu strain calculated from the vector Δu, see Eq. (7.2);

$\tilde{D}$ is the time-dependent material elastic matrix so that $\Delta S = \tilde{D}\,\Delta e$ with ΔS being the incremental second Piola-Kirchhoff stress vector, see Sec. 7.5;

σ is the Cauchy stress vector; and

$W^{t+\Delta t}$ is the work-equivalent term corresponding to prescribed body-force and surface traction in configuration $C^{t+\Delta t}$.

A note on notation may be in order. Although $W^{t+\Delta t}$ which is associated with configuration $C^{t+\Delta t}$ or time "$t+\Delta t$" is superscripted, vectors and matrices such as Δu, Δe, σ, and $\tilde{D}$ that are associated with time "t" are not for conciseness. Variables without superscript "t" are understood to be those at time "t". The exceptions are V^t which is the reference volume and A^t, the mid-surface area of configuration C^t. Reference volume V^t should not be confused with V_i, which is the director of node i and at time "t", see Eq. (7.13), for instance.

In Eq. (7.1), the incremental Washizu strain $\Delta\bar{e}$ is, in component form

$$\Delta\bar{e}_{ij} = \Delta\varepsilon_{ij} + \Delta\eta_{ij}, \quad \Delta\varepsilon_{ij} = \tfrac{1}{2}\left(\Delta u_{i,j} + \Delta u_{j,i}\right), \quad \Delta\eta_{ij} = \tfrac{1}{2}\Delta u_{k,i}\Delta u_{k,j} \dots \tag{7.2}$$

where the Einstein summation convention for indices has been adopted for k and the differentiation is with respect to reference coordinates, x_i ($i = 1, 2, 3$) at time "t".

Now, assuming that within every element it can be written as

$$\Delta u = N\Delta q, \quad \Delta e = \tilde{P}\Delta\beta \dots \tag{7.3}$$

where Δq is the vector of incremental nodal displacements, $\Delta\beta$ the vector of incremental strain parameters, and N and $\tilde{P}$ the matrices of displacement interpolation function and strain distribution function. Details of the two matrices will be introduced in Sec. 7.3. Substituting Eq. (7.3) into (7.1) gives, with V_e being the volume of an element in the reference configuration

$$\Delta\pi_{HR}\left(\Delta u, \Delta e\right) = \Delta\pi_{HR}\left(\Delta q, \Delta\beta\right)$$

$$= \sum \int_{V_e}\left[-\tfrac{1}{2}(\Delta\beta)^T\,\tilde{P}^T\tilde{D}\tilde{P}(\Delta\beta)\right]dV + \sum \int_{V_e}\left[\sigma^T B_L(\Delta q)\right]dV$$

$$+ \sum \int_{V_e}\left[\tfrac{1}{2}(\Delta q)^T\,B_{NL}^T\bar{\sigma}^T B_{NL}(\Delta q)\right]dV + \sum \int_{V_e}\left[(\Delta\beta)^T\,\tilde{P}^T\tilde{D}B_L(\Delta q)\right]dV \cdots \tag{7.4}$$

$$+ \sum \int_{V_e}\left[(\Delta e)^T\,\tilde{D}(\Delta\eta)\right]dV - W^{t+\Delta t}$$

Note that in arriving at Eq. (7.4) it has been defined that

$$\Delta\varepsilon = B_L\left(\Delta q\right), \quad \sigma^T\left(\Delta\eta\right) = \tfrac{1}{2}\left(\Delta q\right)^T B_{NL}^T\bar{\sigma}^T B_{NL}\left(\Delta q\right)... \tag{7.5}$$

where B_L and B_{NL} are the linear and nonlinear strain-displacement matrices that will be discussed in Sec. 7.3, and $\bar{\sigma}$ is a matrix that contains the Cauchy stress components at time "t". The construction of $\bar{\sigma}$ from σ will be given in Sub-sec. 7.3.5. The underlined term on the right-hand side (RHS) of Eq. (7.4) contains the third-order product of Δq and $\Delta\beta$, and will therefore be disregarded when linearizing Eq. (7.1) or (7.4). Next, defining

$$\tilde{H} = \int_{V_e}\left(\tilde{P}^T\tilde{D}\tilde{P}\right)dV, \quad G_e = \int_{V_e}\left(\tilde{P}^T\tilde{D}B_L\right)dV,$$
$$k_{NL} = \int_{V_e}\left(B_{NL}^T\bar{\sigma}B_{NL}\right)dV, \quad f_1 = \int_{V_e}\left(B_L^T\sigma\right)dV \qquad ... \tag{7.6 a-d}$$

and substituting Eq. (7.6) into (7.4) leads to

$$\Delta\pi_{HR}\left(\Delta q,\Delta\beta\right) = \sum -\tfrac{1}{2}\left(\Delta\beta\right)^T\tilde{H}\left(\Delta\beta\right) + \sum f_1\left(\Delta q\right)$$
$$+ \sum\tfrac{1}{2}\left(\Delta q\right)^T k_{NL}\left(\Delta q\right) + \sum\left(\Delta\beta\right)G_e\left(\Delta q\right) - \sum\left[f_e^{t+\Delta t}\right]^T\left(\Delta q\right) \qquad ... \tag{7.7}$$

with $f_e^{t+\Delta t}$ being the external nodal force vector, in configuration $C^{t+\Delta t}$, associated with the $W^{t+\Delta t}$ term of Eq. (7.4). Seeking stationarity of Eq. (7.7) with respect to $\Delta\beta$ yields

$$\Delta\beta = \tilde{H}^{-1}G_e\left(\Delta q\right)... \tag{7.8}$$

Substituting Eq. (7.8) into (7.7) and seeking stationarity with respect to Δq result in the following "equilibrium equation":

$$\left(k_L + k_{NL}\right)\left(\Delta q\right) = f_e^{t+\Delta t} - f_1... \tag{7.9}$$

where

$$k_L = G_e^T\tilde{H}^{-1}G_e ... \tag{7.10}$$

is the element linear or small displacement stiffness matrix. The matrix k_{NL}, as defined by Eq. (7.6), is the element initial stress stiffness matrix. On the RHS of Eq. (7.9), f_1 is the pseudo-force vector. Rewriting $f_e^{t+\Delta t}$-f_1 as $\Delta f + f_e^t$ - f_1, it can be seen that the term $f_e^{t+\Delta t}$-f_1 consists of, Δf, the incremental external force from "t" to "$t+\Delta t$", as well as f_e^t-f_1, the equilibrium imbalance at time "t". If the equilibrium at time "t" is satisfied in an average sense, f_e^t-f_1 vanishes and $f_e^{t+\Delta t}$-f_1 reduces to the increment of external force from "t" to "$t+\Delta t$", that is, Δf.

The above element level formulas are applied to every element. Once all the element matrices are determined they are assembled to form the global "equilibrium equation". This equation is then solved for the displacement increments ΔQ.

Finally, it should be pointed out that, the incremental Washizu strain can be recovered from $\Delta \beta$ through

$$\Delta e = \tilde{P}\left(\Delta \beta\right) = \tilde{P}\tilde{H}^{-1}G_e\left(\Delta q\right)\ldots \tag{7.11}$$

and the incremental second Piola-Kirchhoff stress can be found by

$$\Delta S = \tilde{D}\left(\Delta e\right)\ldots \tag{7.12}$$

7.3. ELEMENT STIFFNESS MATRICES

This section is concerned with the derivation of element stiffness matrices. It contains five sub-sections. Sub-sec. 7.3.1 introduces the element geometry and coordinate systems. Sub-sec. 7.3.2 deals with the assumed incremental displacement field within an element while Sub-sec. 7.3.3 considers the assumed incremental strain field within an element. Derivation of the linear stiffness matrix is included in Sub-sec. 7.3.4 whereas derivation of the initial stress stiffness matrix is presented in Sub-sec. 7.3.5.

7.3.1. Element Geometry and Coordinate Systems

The finite elements under consideration are three node flat shell elements with triangular geometry, as shown in Fig. **7.1**. The three nodes are allocated at the three corners of the mid-surface of the element. A local rectangular coordinate system is attached to node 1, with its r-axis coinciding with the side 1-2, its t-axis being parallel to the normal of the element and its s-axis perpendicular to the r-t plane. With such a coordinate system, the r and s coordinates of nodes 1, 2 and 3 are, (0, 0), (r_2, 0) and (r_3, s_3), respectively. Also defined is the director orthogonal frame, V_r, V_s and V, at any points on the mid-surface. In the undeformed configuration, the director V coincides with the normal of the shell. However, as the shell deforms, the director is, in general, not normal to the mid-surface. Consequently, the director orthogonal frame differs from point to point and from the r-s-t system. Note that setting up the two coordinate systems for every element is necessary in order to capture the large extent to which the shell deforms, especially for deformations of large strain and large rotation. The director orthogonal frame serves as a basis of measuring and recording the change of orientation of points situated on the mid-surface.

The six nodal degrees-of-freedom (DOF) are,

u: displacement in the r-direction,

v: displacement in the s-direction,

w: displacement in the t-direction,

θ_r: rotation component along the r-axis,

θ_s: rotation component along the s-axis, and

θ_t: rotation component along the t-axis, or the drilling degree-of-freedom (DDOF).

The displacements are considered positive if along the positive directions of corresponding axes. For rotations, the right-hand screw rule is adopted in determining their directions. They are considered positive if along the positive directions of corresponding axes.

7.3.2. Assumed Incremental Displacement Field within an Element

Following the isoparametric approach, the local coordinates (r,s,t) of an arbitrary point within the element is

$$\begin{Bmatrix} r \\ s \\ t \end{Bmatrix} = \sum_{i=1}^{3} \xi_i \begin{Bmatrix} r_{0i} \\ s_{0i} \\ 0 \end{Bmatrix} + \eta \sum_{i=1}^{3} \xi_i V_i \quad \dots \tag{7.13}$$

where the subscript "0" that is used with variables r_0 and s_0 indicates that they are defined on the mid-surface. In the remainder of the chapter, mid-surface translational displacements and their increments will also be given the subscript "0". On the other hand, rotational displacements and directors and their increments are understood to be defined on the mid-surface, and are not subscripted with "0". V_i (i = 1, 2, 3) denotes the director of node i at time "t". Finally, ξ_i (i = 1, 2, 3) is the natural or area coordinates of a triangle satisfying

$$0 \le \xi_i \le 1, \quad \sum_{i=1}^{3} \xi_i = 1 \dots \tag{7.14 a,b}$$

and η is the coordinate along the director direction and satisfies

$$-\tfrac{1}{2} h \le \eta \le \tfrac{1}{2} h \dots \tag{7.1 4c}$$

In Eq. (7.14c), h is the thickness of the shell at time "t", which is considered constant over the entire element in the present study. Note that the first summation in Eq. (7.13) represents the position of the mid-surface while the second summation indicates that the director orthogonal frame is interpolated in exactly the same way as the mid-surface r and s coordinates.

The incremental displacements of any point within the element from time "t" to "$t+\Delta t$" can be expressed as

$$\begin{Bmatrix} \Delta u \\ \Delta v \\ \Delta w \end{Bmatrix} = \sum_{i=1}^{3} \xi_i \begin{Bmatrix} \Delta u_{0i} \\ \Delta v_{0i} \\ \Delta w_{0i} \end{Bmatrix} + \eta \sum_{i=1}^{3} \xi_i \Delta V_i \quad \dots \tag{7.15}$$

In Eq. (7.15) the first summation represents the incremental displacements of any points located on the mid-surface. The second summation reflects the change in orientation of director at any points on the mid-surface which is interpolated from ΔV_i (i = 1, 2, 3), the increment of director of node i from time "t" to "$t+\Delta t$". The incremental director, ΔV_i can be expressed as [7.3, 7.4],

$$\Delta V_i = \Delta \theta_i' \times V_i = -V_i \times \Delta \theta_i' = -\Omega_i \Delta \theta_i' \dots \tag{7.16 a}$$

where "$\times$" denotes cross product, $\Delta \theta_i'$ is the incremental rotational vector of node i from time "t" to time "$t+\Delta t$", relative to the director orthogonal frame attached to the same node. The skew-symmetric matrix Ω_i is associated with V_i of node i and time "t" and is defined as,

$$\Omega_i = \begin{bmatrix} 0 & -V_{ti} & V_{si} \\ V_{ti} & 0 & -V_{ri} \\ -V_{si} & V_{ri} & 0 \end{bmatrix} \dots \tag{7.16 b}$$

where V_{ri}, V_{si}, and V_{ti} are the components of V_i. That is, $V_i = [V_{ri}, V_{si}, V_{ti}]^T$.

It is important to recognize that V_i is generally not coincident with the t-axis. Therefore, $\Delta \theta_i'$ which is defined with respect to V_i is not the same as $\Delta \theta_i$, the incremental rotational vector relative to the t-axis. The transformation between $\Delta \theta_i'$ and $\Delta \theta_i$ is the so-called exponential mapping. According to [7.3, 7.4] it can be written as

$$\Delta \theta_i' = \Gamma_i \Delta \theta_i \dots \tag{7.17 a}$$

where Γ_i is an orthogonal matrix associated with node i at time "t". It can be determined by [7.3-7.4],

$$\Gamma_i = aI_3 + \hat{B}_i + \frac{1}{1+a} B_i \hat{B}_i \ldots \qquad (7.17\ \text{b})$$

with I_3 being the 3×3 identity matrix, $a = V_{ti}$, $B_i = \begin{bmatrix} 0 & 0 & 1 \end{bmatrix}^T \times V_i = \begin{bmatrix} V_{si} & V_{ri} & 0 \end{bmatrix}^T$ and $\hat{B}_i$ is a 3×3 skew-symmetric matrix constructed from vector B_i. This matrix is,

$$\hat{B}_i = \begin{bmatrix} & & V_{ri} \\ & & -V_{si} \\ -V_{ri} & V_{si} & \end{bmatrix} \ldots \qquad (7.17\ \text{c})$$

Next, substituting Eqs. (7.16a) and (7.17a) into (7.15), and disregarding the DDOF lead to

$$\begin{Bmatrix} \Delta u \\ \Delta v \\ \Delta w \end{Bmatrix} = \sum_{i=1}^{3} \xi_i \begin{Bmatrix} \Delta u_{0i} \\ \Delta v_{0i} \\ \Delta w_{0i} \end{Bmatrix} + \eta \sum_{i=1}^{3} \xi_i \begin{bmatrix} \Lambda_{i11} & \Lambda_{i12} \\ \Lambda_{i21} & \Lambda_{i22} \\ \Lambda_{i31} & \Lambda_{i32} \end{bmatrix} \begin{Bmatrix} \Delta\theta_{ri} \\ \Delta\theta_{si} \end{Bmatrix} \ldots \qquad (7.18)$$

where Λ_{imn} is the element located in the *m*-th row and *n*-th column of Λ_i which is defined as

$$\Lambda_i = -\Omega_i \overline{\Gamma}_i \ldots \qquad (7.19)$$

with $\overline{\Gamma}_i$ consisting of the first two columns of Γ_i. Since both Ω_i and Γ_i are dependent of the director V_i, Λ_i also depends on the current position of the director. As will be seen later, Eqs. (7.22) and (7.23) in particular, the interpolation of incremental displacements requires Λ_i or every node at every time step. Therefore, it is necessary to update the director at every time step. This is discussed in Sub-sec. 7.6.1 as part of configuration updating.

By forming Λ_i through Eq. (7.19), the DDOF are excluded to ensure the preservation of the normality condition $\Delta\theta_i' \cdot V_i = 0$. To include DDOF, as in Chapter 2, Allman's formula [7.5] of incorporating DDOF with in-plane displacements *u* and *v*, and Tessler and Hughes's approach [7.6] of coupling lateral displacement *w* with the two bending rotations θ_r and θ_s are introduced so that Eq. (7.18) is extended to

$$\begin{Bmatrix} \Delta u \\ \Delta v \\ \Delta w \end{Bmatrix} = \sum_{i=1}^{3} \xi_i \begin{Bmatrix} \Delta u_{0i} \\ \Delta v_{0i} \\ \Delta w_{0i} \end{Bmatrix} + \eta \sum_{i=1}^{3} \xi_i \begin{bmatrix} \Lambda_{i11} & \Lambda_{i12} \\ \Lambda_{i21} & \Lambda_{i22} \\ \Lambda_{i31} & \Lambda_{i32} \end{bmatrix} \begin{Bmatrix} \Delta \theta_{ri} \\ \Delta \theta_{si} \end{Bmatrix} + \sum_{i=1}^{3} \begin{bmatrix} & & \overline{p}_i \\ & & \overline{q}_i \\ -\overline{p}_i & -\overline{q}_i & \end{bmatrix} \begin{Bmatrix} \Delta \theta_{ri} \\ \Delta \theta_{si} \\ \Delta \theta_{ti} \end{Bmatrix} \dots \ (7.20)$$

where the terms $\overline{p}_i$ and $\overline{q}_i$ are defined by Eq. (2.20), and the a_{ij} and b_{ij} terms are given by Eq. (2.21). Equation (7.20) can also be recast into a more compact form

$$\begin{bmatrix} \Delta u & \Delta v & \Delta w \end{bmatrix}^T =$$

$$N \begin{bmatrix} \Delta u_{01} & \Delta v_{01} & \Delta w_{01} & \Delta \theta_{r1} & \Delta \theta_{s1} & \Delta \theta_{t1} & \Delta u_{02} & \dots & \Delta \theta_{s3} & \Delta \theta_{t3} \end{bmatrix}^T \ \dots \qquad (7.21)$$

where

$$N = \begin{bmatrix} N_1 & N_2 & N_3 \end{bmatrix} \dots \qquad (7.22 \ a)$$

and the 3×6 sub-matrix N_i is

$$N_i = \begin{bmatrix} \xi_i & & \eta \xi_i \Lambda_{i11} & \eta \xi_i \Lambda_{i12} & \overline{p}_i \\ & \xi_i & \eta \xi_i \Lambda_{i21} & \eta \xi_i \Lambda_{i22} & \overline{q}_i \\ & & \xi_i & \eta \xi_i \Lambda_{i31} - \overline{p}_i & \eta \xi_i \Lambda_{i32} - \overline{q}_i & \end{bmatrix} \dots \qquad (7.22 \ b)$$

The interpolation function for DDOF, which is needed for the derivation of the present element but not included in usual isoparametric formulations, is

$$\Delta \theta_t = \sum_{i=1}^{3} \xi_i \Delta \theta_{ti} \ \dots \qquad (7.22 \ c)$$

Note that DDOF are defined on the mid-surface of the shell.

7.3.3. Assumed Incremental Strain Field within an Element

Similar to Chapter 2, the assumed incremental strain field for any point within the element is

$$\begin{bmatrix} \Delta \varepsilon_r & \Delta \varepsilon_s & \Delta \varepsilon_{rs} & \Delta \varepsilon_{st} & \Delta \varepsilon_{tr} \end{bmatrix}^T = \tilde{P} \begin{bmatrix} \Delta \beta_1 & \Delta \beta_2 & \Delta \beta_3 & \Delta \beta_4 & \Delta \beta_5 & \dots & \Delta \beta_9 \end{bmatrix}^T \dots \ (7.23)$$

where

$$\tilde{P} = \begin{bmatrix} 1 & & & \eta & & & \\ & 1 & & & \eta & & \\ & & 1 & & & \eta & \\ & & & s_3 z_2 & s_3\left(z_2 + 2\xi_3\right) & \\ & & & r_3 z_2 & r_{32}\left(z_2 + 2\xi_3\right) & -r_2 z_3 \end{bmatrix} \cdots \qquad (7.24)$$

with $r_{32} = r_3\text{-}r_2$, $z_2 = 2\xi_2\text{-}1$ and $z_3 = 2\xi_3\text{-}1$. Note that the assumed incremental strain components $\Delta\varepsilon_{rs}$, $\Delta\varepsilon_{st}$, and $\Delta\varepsilon_{tr}$ use the so-called engineering definition. Among the strain parameters, $\Delta\beta_1$ through $\Delta\beta_3$ are associated with membrane strains, $\Delta\beta_4$ through $\Delta\beta_6$ bending strains, and $\Delta\beta_7$ through $\Delta\beta_9$ transversal strains.

7.3.4. Element Linear Stiffness Matrix k_L

Section 7.2 showed that at any time instant the total element stiffness matrix is the sum of the linear and initial stress stiffness matrices, k_L and k_{NL}. The derivation of the element linear stiffness matrix k_L is discussed in this sub-section while the element initial stress stiffness matrix k_{NL} will be dealt with in the next sub-section.

The derivatives of the incremental displacements, Eq. (7.21), with respect to local co-ordinates r, s and t, are

$$\begin{aligned} &\begin{bmatrix} \Delta u_{,r} & \Delta v_{,r} & \Delta w_{,r} & \Delta u_{,s} & \Delta v_{,s} & \Delta w_{,s} & \Delta u_{,t} & \Delta v_{,t} & \Delta w_{,t} \end{bmatrix}^T \\ &= \sum_{i=1}^{3} H_i \begin{bmatrix} \Delta u_{0i} & \Delta v_{0i} & \Delta w_{0i} & \Delta\theta_{ri} & \Delta\theta_{si} & \Delta\theta_{ti} \end{bmatrix}^T \quad \cdots \end{aligned} \qquad (7.25)$$

where H_i ($i = 1, 2, 3$) is a 9×6 matrix defined as

$$H_i = \begin{bmatrix} H_{i(r)} \\ H_{i(s)} \\ H_{i(t)} \end{bmatrix} \cdots \qquad (7.26)$$

with sub-matrices of

$$H_{i(r)} = \begin{bmatrix} \xi_{i,r} & & \eta\xi_{i,r}\Lambda_{i11} & \eta\xi_{i,r}\Lambda_{i12} & \overline{p}_{i,r} \\ & \xi_{i,r} & \eta\xi_{i,r}\Lambda_{i21} & \eta\xi_{i,r}\Lambda_{i22} & \overline{q}_{i,r} \\ & \xi_{i,r} & \eta\xi_{i,r}\Lambda_{i31} - \overline{p}_{i,r} & \eta\xi_{i,r}\Lambda_{i32} - \overline{q}_{i,r} & \end{bmatrix} \dots \quad (7.27\ a)$$

and

$$H_{i(t)} = \begin{bmatrix} 0 & 0 & 0 & \xi_i\Lambda_{i11} & \xi_i\Lambda_{i21} & 0 \\ & & & \xi_i\Lambda_{i21} & \xi_i\Lambda_{i22} & \\ & & & \xi_i\Lambda_{i31} & \xi_i\Lambda_{i32} & \end{bmatrix} \dots \quad (7.27\ b)$$

The sub-matrix $H_{i(s)}$ is similar in form to Eq. (7.27a) except that the derivatives with respect to r in Eq. (7.27a) are replaced by those with respect to s. The derivatives of ξ_i ($i = 1, 2, 3$) with respect to r and s are

$$\xi_{1,r} = -\frac{1}{r_2} \ , \ \xi_{1,s} = \frac{r_3 - r_2}{r_2 s_3} \ , \ \xi_{2,r} = \frac{1}{r_2} \ , \ \xi_{2,s} = -\frac{r_3}{r_2 s_3} \ , \ \xi_{3,r} = 0 \ , \ \xi_{3,s} = \frac{1}{s_3} \dots \quad (7.28)$$

The incremental strain-displacement relationships for any points within the element, using engineering definition of strains, are

$$\Delta\varepsilon_r = \Delta u,_r \ , \ \Delta\varepsilon_s = \Delta v,_s \ , \ \Delta\varepsilon_{rs} = \Delta u,_s + \Delta v,_r \ ,$$
$$\Delta\varepsilon_{st} = \Delta v,_t + \Delta w,_s \ , \ \Delta\varepsilon_{tr} = \Delta w,_r + \Delta u,_t \qquad \dots \qquad (7.29)$$

or in a more compact form

$$\begin{bmatrix} \Delta\varepsilon_r & \Delta\varepsilon_s & \Delta\varepsilon_{rs} & \Delta\varepsilon_{st} & \Delta\varepsilon_{tr} \end{bmatrix}^T = \sum_{i=1}^{3} I_I H_i \begin{bmatrix} \Delta u_{0i} & \Delta v_{0i} & \Delta w_{0i} & \Delta\theta_{ri} & \Delta\theta_{si} & \Delta\theta_{ti} \end{bmatrix}^T \dots (7.30)$$

with the 5×9 matrix I_I being

$$I_I = \begin{bmatrix} 1 & & & & & & & & 0 \\ & & 1 & & & & & & \\ 1 & & 1 & & & & & & \\ & & & & 1 & & 1 & & \\ & 1 & & & & & & 1 & \end{bmatrix} \dots \quad (7.31)$$

and H_i has been defined in Eq. (7.26). Note that $\sum_{i=1}^{3} I_I H_i = B_L$ is the linear strain-displacement matrix B_L.

Now, the element stiffness matrix k_H which takes into account the effects due to membrane, bending, and transversal shear can be obtained by applying Eqs. (7.6a-b) and (7.10). That is,

$$k_H = G_e^T \tilde{H}^{-1} G_e \ldots \tag{7.32}$$

where $\tilde{H}$ and G_e are defined in Eq. (2.6).

Similar to Chapter 2, the effect due to the DDOF is accounted for by including k_t which is defined by Eqs. (2.34) and (2.35) so that k_L becomes

$$k_L = k_H + k_t \ldots \tag{7.33}$$

It should be mentioned that the full integral of Eq. (2.34) results in an element that is able to provide the correct number of rigid body modes. On the other hand, using one-point quadrature to evaluate Eq. (2.34) yields better results than using the full integration [7.7-7.8]. This one-point quadrature can be easily achieved by setting $\xi_1 = \xi_2 = \xi_3 = 1/3$ in Eq. (2.35). As pointed out in Sub-sec. 2.2.8 this element, however, has two spurious modes.

7.3.5. Element Initial Stress Stiffness Matrix k_{NL}

To obtain the element initial stress stiffness matrix k_{NL} one first writes the nonlinear strain-displacement matrix B_{NL}

$$B_{NL} = \begin{bmatrix} H_1 & H_2 & H_3 \end{bmatrix} \ldots \tag{7.34}$$

where the 9×6 matrices H_i ($i = 1, 2, 3$) have been defined in Eqs. (7.26) and (7.27a-b).

The evaluation of k_{NL} also requires the Cauchy or true stress matrix $\bar{\sigma}$ which is constructed from the Cauchy stress vector $\sigma = [\sigma_r, \sigma_s, \sigma_{rs}, \sigma_{st}, \sigma_{tr}]^T = [\sigma_{11}, \sigma_{22}, \sigma_{12}, \sigma_{23}, \sigma_{31}]^T$, according to the following formula

$$\bar{\sigma} = \begin{bmatrix} \sigma_{11}I_3 & \sigma_{12}I_3 & \sigma_{31}I_3 \\ \sigma_{12}I_3 & \sigma_{22}I_3 & \sigma_{23}I_3 \\ \sigma_{31}I_3 & \sigma_{23}I_3 & O_3 \end{bmatrix} \dots \tag{7.35}$$

with I_3 being the 3×3 identity matrix and O_3 a 3×3 null matrix.

Equations (7.11), (7.12) and (7.24) indicate that, the membrane components of the incremental second Piola-Kirchhoff stress, ΔS, are constant within an element. The bending components vary in the thickness direction only. However, the transverse components are mid-surface position dependent, or more precisely, are varied linearly within an element. Consequently, S, the second Piola-Kirchhoff stress at time "t" and σ, the Cauchy stresses at time "t", are also a function of mid-surface position. Their updating requires storage of stresses at the three nodes. Returning to the integration of Eq. (7.6), if σ is considered mid-surface position dependent, it would produce very tedious expressions. Therefore, in the present study the transverse stress components of σ are considered constant over an element. That is, all the stress components of σ are calculated and updated only at the centroid of every element. This approach can be justified by recognizing that in shell structures membrane and bending stresses are usually dominant. Such an approach reduces substantially the computation efforts involved and requires much less computer memory.

In general, every σ_{ij} in Eq. (7.35) is a combination of membrane, bending and transverse components. At any point, including the centroid of the element, membrane and transverse components can be obtained by setting η to zero for the purpose of evaluating k_{NL}. The stress at the top surface can be obtained by prescribing $\eta = h/2$. The difference between the stresses at the top surface and the mid-surface, after divided by $h/2$, indicates the slope of bending stress component. That is, σ_{ij} can be re-written as

$$\sigma_{ij} = \sigma_{ij(0)} + \eta d\sigma_{ij}, \quad d\sigma_{ij} = \frac{2}{h}\left(\sigma_{ij(+)} - \sigma_{ij(0)}\right) \dots \tag{7.36 a-b}$$

where $\sigma_{ij(0)}$ and $\sigma_{ij(+)}$ denote stresses at the middle and top surfaces, respectively, while $d\sigma_{ij}$ is the slope of the bending stress. Equation (7.36) is also valid for plastic deformation. In this case, $d\sigma_{ij}$ becomes zero because of the adoption of the non-layered approach to be introduced in Sec. 7.5 below. Of course, for layered approach $d\sigma_{ij}$ is generally not zero.

With the Cauchy stress matrix $\bar{\sigma}$ and B_{NL} matrix now defined, they are substituted into Eq. (7.6c) so that explicit expressions for k_{NL} can be obtained after evaluating the required integration. Expressing σ_{ij} by Eq. (7.36) makes the explicit expressions possible and manageable.

In passing, it should be mentioned that the above derivation will result in the director version of the element linear stiffness matrix and the element initial stress stiffness matrix. The simplified version, on the other hand, can be obtained by setting the directors at all nodes to $V_i = [0\ 0\ 1]^T$. As a result, the Γ_i matrix of Eq. (7.17b) becomes I_3, the 3×3 identity matrix, and the Λ_i matrix of Eq. (7.19) becomes

$$\Lambda_i = - \begin{bmatrix} 0 & -1 \\ 1 & 0 \\ 0 & 0 \end{bmatrix} \dots \tag{7.37}$$

Finally, the simplified version of the element linear stiffness matrix coincides with that developed in Chapter 2, Sub-sec. 2.2.7 and 2.2.8 in particular.

7.4. CONSISTENT ELEMENT MASS MATRICES

The scope of the present study is limited to cases in which angular velocities and accelerations are small and can therefore be disregarded when formulating the consistent element mass matrices. In this section it will be shown that, under such conditions, the element mass matrix that was derived by Simo *et al.* [7.9] using the conserving algorithm reduces to that presented in Sub-sec. 2.2.9.

As Eq. (7.13) indicated, the local coordinates of a point within an element can be determined by its mid-surface position vector, the director and the distance of the point to the mid-surface. Equation (7.13) can therefore be recast as

$$\Phi = \varphi + \eta V \,\ldots \tag{7.38}$$

where Φ denotes the LHS of Eq. (7.13) while φ and ηV are the first and second summation terms on the RHS of Eq. (7.13), respectively. For points on the mid-surface, the linear momentum field is defined as, with ρ being the mass density at time "t"

$$p = \int_{-\frac{h}{2}}^{\frac{h}{2}} \rho \dot{\Phi} d\eta = \int_{-\frac{h}{2}}^{\frac{h}{2}} \rho \dot{\varphi} d\eta = \rho h \dot{\varphi} \,\ldots \tag{7.39 a}$$

and the angular momentum field is

$$\chi = \int_{-\frac{h}{2}}^{\frac{h}{2}} \rho(\Phi - \varphi) \times \dot{\Phi} d\eta = \int_{-\frac{h}{2}}^{\frac{h}{2}} \rho \left[\eta V \times \dot{\varphi} + \eta^2 V \times \dot{V} \right] d\eta$$

$$\ldots \tag{7.39 b}$$

$$= \int_{-\frac{h}{2}}^{\frac{h}{2}} \rho \left[\eta^2 V \times \dot{V} \right] d\eta = \rho \frac{h^3}{12} V \times \dot{V} = I_\rho V \times \dot{V}$$

where the integral $\int_{-\frac{h}{2}}^{\frac{h}{2}} \eta d\eta = 0$ has been applied and I_ρ is,

$$I_\rho = \rho \frac{h^3}{12} \,\ldots \tag{7.39 c}$$

Considering the weak form of momentum balance and defining virtual mid-surface displacement $\Delta\varphi$ and virtual rotation of the director field $\Delta\alpha$ ($\Delta\varphi$ and $\Delta\alpha$ are considered constant over the time period "t" to "$t+\Delta t$"), one writes [7.9]

$$G_{dyn}(\Phi, \Delta\varphi, \Delta\alpha) = \int_{A^t} \left[(\Delta\varphi)^T \dot{p} + (\Delta\alpha)^T \dot{\chi} \right] dA + G_{sta}(\Phi, \Delta\varphi, \Delta\alpha) = 0 \,\ldots \tag{7.40}$$

In Eq. (7.43), A^t is the mid-surface area of the reference configuration, G_{sta} is the static part of the weak form and will give the stiffness matrices, pseudo-force and consistent external load vectors. In the present study, G_{sta} is in fact the linearized incremental variational principle discussed in Sec. 7.2. The integral in Eq. (7.40) represents the effects due to the linear and angular velocities and accelerations.

This integral comprises of two parts, one due to translational motion and the other rotational motion. Writing the integral as

$$\int_{A'} \left[(\Delta\varphi)^T \, \dot{p} + (\Delta\alpha)^T \, \dot{\chi} \right] dA = G_{tra} + G_{rot} = \int_{A'} (\Delta\varphi)^T \, \dot{p} dA + \int_{A'} (\Delta\alpha)^T \, \dot{\chi} dA \dots \quad (7.41)$$

with

$$\dot{p} = \rho h \ddot{\varphi} \, , \qquad \dot{\chi} = I_\rho \left(\dot{V} \times \dot{V} + V \times \ddot{V} \right) = I_\rho V \times \ddot{V} \dots \qquad (7.42 \text{ a-b})$$

For G_{tra}, taking its variation with respect to $\dot{p}$ leads to

$$\Delta G_{tra} = \int_{A'} (\Delta\varphi)^T \, \Delta\dot{p} dA = \int_{A'} (\Delta\varphi)^T \, \rho h \Delta\ddot{\varphi} dA = \sum (\Delta q) \left\{ \int_{A_e} \rho h N_m^{\ T} N_m dA \right\} (\Delta\ddot{q}) \dots \quad (7.43)$$

where Δq is the vector of nodal DOF and $\Delta\ddot{q}$ is the vector of acceleration of the nodal DOF. In arriving at Eq. (7.43), the relations

$$\Delta\varphi = N_m \Delta q, \qquad \Delta\ddot{\varphi} = N_m \Delta\ddot{q} \dots \qquad (7.44 \text{ a-b})$$

have been applied, where N_m is the interpolation function matrix for mid-surface displacements. The component matrix of N_m is N_{mi} ($i = 1, 2, 3$). The latter can be obtained by setting to zero the director direction coordinate η in Eq. (7.22b) and adding three rows of zeros for consistency of matrix dimensions. In other words,

$$N_{mi} = \begin{bmatrix} \xi_i & & & & \bar{p}_i \\ & \xi_i & & & \bar{q}_i \\ & & \xi_i & -\bar{p}_i & -\bar{q}_i \\ 0 & & & & \\ 0 & & & & \\ 0 & & & & \end{bmatrix} \dots \qquad (7.44 \text{ c})$$

Note that Eq. (7.43) has in fact defined the translational part of the element consistent mass matrix m_{tra}:

$$m_{tra} = \int_{A_e} \rho h N_m^{\ T} N_m dA \ldots \tag{7.45}$$

Now, considering the G_{rot} in Eq. (7.41) and taking its variation with respect to V and $\ddot{V}$, one can show that

$$\Delta G_{rot} = \int_{A^t} I_\rho (\Delta\alpha)^T (\Delta V \times \ddot{V} + V \times \Delta \ddot{V}) dA \approx \int_{A^t} I_\rho (\Delta\alpha)^T (V \times \Delta \ddot{V}) dA \ldots \tag{7.46}$$

in which $\ddot{V}$ represents the angular accelerations since the directors are assumed to be of unit length. Note that the term $\Delta V \times \ddot{V}$ has been disregarded because of the small angular velocity and small acceleration assumptions. By making use of Eq. (7.16a) and interchanging the sequence of variation and differentiation, it results in

$$\Delta \ddot{V} = (\Delta \ddot{\alpha} \times V + \Delta\alpha \times \ddot{V}) \approx \Delta \ddot{\alpha} \times V \ldots \tag{7.47}$$

in which the term associated with the angular accelerations has been disregarded because of the small acceleration assumption. Substituting Eq. (7.47) into (7.46), one has

$$\Delta G_{rot} \approx \int_{A^t} I_\rho (\Delta\alpha)^T V \times (-V \times \Delta \ddot{\alpha}) dA = \int_{A^t} I_\rho (\Delta\alpha)^T (\Delta \ddot{\alpha}) dA$$

$$= \sum (\Delta q)^T \left\{ \int_{A_e} I_\rho N_r^{\ T} N_r dA \right\} (\Delta \ddot{q}) \qquad \ldots \tag{7.48}$$

In deriving Eq. (7.48), it has been assumed that

$$\Delta\alpha = N_r \Delta q, \qquad \Delta \ddot{\alpha} = N_r \Delta \ddot{q} \ldots \tag{7.49 a-b}$$

where N_r and its component matrices

$N_{ri}(i = 1, 2, 3)$ are,

$$N_r = \begin{bmatrix} N_{r1} & N_{r2} & N_{r3} \end{bmatrix}, \qquad N_{ri} = \begin{bmatrix} O_{3\times3} & O_{3\times2} & O_{3\times1} \\ O_{3\times3} & \xi_i \Gamma_i \Lambda_i & O_{3\times1} \end{bmatrix} \ldots \tag{7.49 c-d}$$

with Γ_i being defined by Eq. (7.17b) and Λ_i by Eq. (7.19). In Eq. (7.49d), $O_{m \times n}$ means a $m \times n$ zero matrix.

Again, Eq. (7.48) has defined the rotational part of the element consistent mass matrix which is associated with the bending rotational DOF. That is,

$$m_{rot} = \int_{A_e} I_\rho N_r^T N_r dA \ldots \tag{7.50}$$

What remains to be considered is the part of the element consistent mass matrix that corresponds to the DDOF. This part of the matrix is

$$m_{ddof} = \int_{A_e} J_d N_d^T N_d dA \ldots \tag{7.51}$$

where the component matrices of N_d is

$$N_{di} = \begin{bmatrix} 0 & 0 & 0 & 0 & 0 & \xi_i \end{bmatrix} \ldots \tag{7.52 a}$$

and

$$J_d = \rho \frac{r_2^2 + s_3^2 + r_3 \left(r_3 - r_2 \right)}{18} \ldots \tag{7.52 b}$$

The consistent element mass matrix m is then,

$$m = m_{tra} + m_{rot} + m_{ddof} \ldots \tag{7.53}$$

Similar to the element stiffness matrices, the simplified version of the consistent element mass matrix can be obtained by setting the Λ_i matrix to that of Eq. (7.37). This simplified version of consistent element mass matrix is proven to be the same as that shown in Sub-sec. 2.2.9.

In passing, it should be pointed out that the consistent element mass matrix has to be updated at every time step since the mass density, thickness of the shell and element geometry change from time to time. The approach of updating mass matrix, instead of forming and using only the initial mass matrices, will be employed in the next two chapters in which numerical examples of nonlinear dynamic analysis will be presented.

7.5. CONSTITUTIVE RELATIONS

This section is concerned with the constitutive relations for linear, elastic and isotropic materials with small or finite strains, and elasto-plastic materials with isotropic strain hardening. The latter case also includes deformations of small or finite strains.

7.5.1. Elastic Materials

The constitutive relation for homogeneous, isotropic and linearly elastic materials of small strain is

$$
D = \begin{bmatrix} a & b & b & & & \\ b & a & b & & & \\ b & b & a & & & \\ & & & f & & \\ & & & & \kappa_s f & \\ & & & & & \kappa_s f \end{bmatrix} \cdots
\tag{7.54}
$$

with

$$
a = \frac{E}{1+v}\frac{1-v}{1-2v} \; , b = \frac{E}{1+v}\frac{v}{1-2v} \; , f = \frac{E}{2(1+v)} \cdots
\tag{7.55 a-c}
$$

where E is the Young's modulus, v the Poisson's ratio and $\kappa_s = 5/6$ the form factor of shear. The stress and strain vectors accompanying Eq. (7.55) are

$$
\sigma = \begin{bmatrix} \sigma_r & \sigma_s & \sigma_t & \sigma_{rs} & \sigma_{st} & \sigma_{tr} \end{bmatrix}^T = \begin{bmatrix} \sigma_{11} & \sigma_{22} & \sigma_{33} & \sigma_{12} & \sigma_{23} & \sigma_{31} \end{bmatrix}^T
$$
$$
\varepsilon = \begin{bmatrix} \varepsilon_r & \varepsilon_s & \varepsilon_t & \varepsilon_{rs} & \varepsilon_{st} & \varepsilon_{tr} \end{bmatrix}^T = \begin{bmatrix} \varepsilon_{11} & \varepsilon_{22} & \varepsilon_{33} & \varepsilon_{12} & \varepsilon_{23} & \varepsilon_{31} \end{bmatrix}^T
\tag{7.56 a-b}
$$

Equation (7.54) should be reduced from general three-dimensional applications to plate or shell analyses by the imposition of the zero normal stress condition, $\sigma_{33} = 0$. Hughes and Liu [7.10] proposed to perform a transformation that is defined as

$$\tilde{D} = T_c^T D T_c \dots \qquad (7.57)$$

where

$$T_c = \begin{bmatrix} 1 & & & & & \\ & 1 & & & & \\ t_1 & t_2 & t_4 & t_5 & t_6 & \\ & & 1 & & & \\ & & & 1 & & \\ & & & & 1 & \end{bmatrix} \dots \qquad (7.58\ a)$$

and

$$t_k = -\frac{D_{3k}}{D_{33}} \dots \qquad (7.58\ b)$$

where D_{3k} is the element located in the third row and k-th column of D.

For finite strain deformations, Ref. [7.11, 7.12] suggested to add to Eq. (7.54) the following term, when cast in matrix form:

$$D' = -\frac{1}{2} \begin{bmatrix} 4\sigma_{11} & & & 2\sigma_{12} & & 2\sigma_{13} \\ & 4\sigma_{22} & & 2\sigma_{12} & 2\sigma_{23} & \\ & & & 2\sigma_{23} & 2\sigma_{13} & \\ 2\sigma_{12} & 2\sigma_{12} & & \sigma_{11}+\sigma_{22} & \sigma_{13} & \sigma_{23} \\ & 2\sigma_{23} & 2\sigma_{23} & \sigma_{13} & \sigma_{22} & \sigma_{12} \\ 2\sigma_{13} & & 2\sigma_{13} & \sigma_{23} & \sigma_{12} & \sigma_{11} \end{bmatrix} \dots \qquad (7.59)$$

Note that D' comes as a result of transforming the Jaumann stress rate to the incremental second Piola-Kirchhoff stress. The transformation rule of Eq. (7.57) also applies to $D+D'$.

7.5.2. Elasto-plastic Materials with Isotropic Strain Hardening

Confining to applications of small elastic, but large plastic strain (thus, large total strain), the J_2 flow theory of plasticity [7.11-7.13] is deemed appropriate. The small strain formulation of the J_2 flow theory involves the stress deviator σ^D and the J_2 invariant of stresses which are, in Cartesian co-ordinates

$$\sigma_{ij}^D = \sigma_{ij} - \frac{1}{3}\sigma_{kk}\delta_{ij} \ , \ J_2 = \frac{1}{2}\sigma_{ij}^D\sigma_{ij}^D \ \dots \tag{7.60 a-b}$$

where δ_{mn} is the Kronecker delta and the Einstein summation convention has been adopted for index k. The constitutive relation in matrix form is then

$$D^{ep} = D + D^{\alpha} \ \dots \tag{7.61}$$

where D is by Eq. (7.54) and D^{α} is,

$$D^{\alpha} = -\beta \begin{bmatrix} \sigma_{11}^D\sigma_{11}^D & \sigma_{11}^D\sigma_{22}^D & \sigma_{11}^D\sigma_{33}^D & \sigma_{11}^D\sigma_{12} & \sigma_{11}^D\sigma_{23} & \sigma_{11}^D\sigma_{31} \\ \sigma_{22}^D\sigma_{11}^D & \sigma_{22}^D\sigma_{22}^D & \sigma_{22}^D\sigma_{33}^D & \sigma_{22}^D\sigma_{12} & \sigma_{22}^D\sigma_{23} & \sigma_{22}^D\sigma_{31} \\ \sigma_{33}^D\sigma_{11}^D & \sigma_{33}^D\sigma_{22}^D & \sigma_{33}^D\sigma_{33}^D & \sigma_{33}^D\sigma_{12} & \sigma_{33}^D\sigma_{23} & \sigma_{33}^D\sigma_{31} \\ \sigma_{12}\sigma_{11}^D & \sigma_{12}\sigma_{22}^D & \sigma_{12}\sigma_{33}^D & \sigma_{12}\sigma_{12} & \sigma_{12}\sigma_{23} & \sigma_{12}\sigma_{31} \\ \sigma_{23}\sigma_{11}^D & \sigma_{23}\sigma_{22}^D & \sigma_{23}\sigma_{33}^D & \sigma_{23}\sigma_{12} & \sigma_{23}\sigma_{23} & \sigma_{23}\sigma_{31} \\ \sigma_{31}\sigma_{11}^D & \sigma_{31}\sigma_{22}^D & \sigma_{31}\sigma_{33}^D & \sigma_{31}\sigma_{12} & \sigma_{31}\sigma_{23} & \sigma_{31}\sigma_{31} \end{bmatrix} \ \dots \tag{7.62 a}$$

with

$$\beta = \frac{\alpha}{\psi} \ , \ \psi = \frac{2}{3}\bar{\sigma}^2 \frac{E - \frac{1-2\nu}{3}E_T}{E - E_T} \dots \tag{7.62 b-c}$$

In Eq. (7.62c), E_T is the tangent modulus of plasticity, E the Young's modulus, ν the Poisson's ratio, and $\bar{\sigma} = \sqrt{3J_2}$ is the effective stress. The parameter α in Eq. (7.62b) has the value of either zero or unity. When $\alpha = 0$ it is associated with elastic loading or any unloading, and when $\alpha = 1$ it is associated with plastic loading.

Two comments are in order. Firstly, it is usually assumed that the effects of transversal shear stresses on plastic behaviours can be disregarded [7.14]. Such an assumption leads to a simplified D^α matrix which becomes now

$$D^\alpha = -\beta \begin{bmatrix} \sigma_{11}^D\sigma_{11}^D & \sigma_{11}^D\sigma_{22}^D & \sigma_{11}^D\sigma_{33}^D & \sigma_{11}^D\sigma_{12} & 0 & 0 \\ \sigma_{22}^D\sigma_{11}^D & \sigma_{22}^D\sigma_{22}^D & \sigma_{22}^D\sigma_{33}^D & \sigma_{22}^D\sigma_{12} & & \\ \sigma_{33}^D\sigma_{11}^D & \sigma_{33}^D\sigma_{22}^D & \sigma_{33}^D\sigma_{33}^D & \sigma_{33}^D\sigma_{12} & & \\ \sigma_{12}\sigma_{11}^D & \sigma_{12}\sigma_{22}^D & \sigma_{12}^D\sigma_{33}^D & \sigma_{12}\sigma_{12} & & \\ 0 & & & & & \\ 0 & & & & & \end{bmatrix} \dots \tag{7.63}$$

Secondly, the transformation rule of Eq. (7.60) is applicable to D^{ep}.

Finally, the above small strain formulation can be extended to finite strain cases in an approach similar to that of Sub-sec. 7.5.1, by adding D' of Eq. (7.59) to D^{ep} of (7.61). That is,

$$D_f^{ep} = D^{ep} + D' = D + D^\alpha + D' \dots \tag{7.64}$$

The transformation rule of Eq. (7.57) should then be applied to D_f^{ep}.

7.5.3. Yield Criterion

By defining the effective stress $\bar{\sigma} = \sqrt{3J_2}$, the von Mises criterion of yield is assumed. The von Mises criterion is applied to metals. It can be expressed in terms of stresses or stress resultants. The first approach allows for the spread of plasticity over the thickness of plates and shells, and is termed "the layered approach". The second approach, "the non-layered approach", on the other hand, employs yield functions that are in terms of stress resultants. This approach assumes that the entire cross-section becomes plastic simultaneously. Therefore, compared with the non-layered approach, the layered approach seems to be more realistic but requires larger amount of algebraic manipulations in forming element stiffness matrix. Ref. [7.14] showed that the discrepancy between the two approaches was insignificant, while [7.15] seemed to suggest employing a large number of elements with the non-layered approach. In the present investigation, the non-layered approach is adopted for two considerations. Firstly, it requires less computation to evaluate element stiffness matrices. Secondly, the D^{ep} or D_f^{ep} matrix can be written in a simple and concise way, which enables one to obtain explicit expressions for the stiffness matrix that are of manageable size.

In the non-layered approach, the yield criterion needs to be expressed in terms of stress resultants. In 1948, Ilyushin applied the von Mises yield criterion to thin shells [7.16]. The idea was further developed by Shapiro [7.17]. Robinson [7.18] had shown that the Ilyushin-Shapiro yield condition can be reduced to a non-parametric form without loss of accuracy and generality. In what follows, the non-parametric form of yield condition investigated by Robinson is introduced, but it is referred to as the Ilyushin's yield condition henceforth.

If S_y is the yield strength of the material in simple tension and h the thickness of the shell, the dimensionless membrane forces n_{11}, n_{22} and n_{12}, and the dimensionless bending moment sm_{11}, m_{22} and m_{12} (note that they should not be confused with the elements of mass matrix) are,

$$n_{ij} = \frac{N_{ij}}{S_y h} \; , \;\; m_{ij} = \frac{4M_{ij}}{S_y h^2} \cdots \tag{7.65 a-b}$$

where the membrane forces N_{ij} and bending moments M_{ij} are related to stress components across the cross-section,

$$N_{ij} = \int_{-\frac{h}{2}}^{\frac{h}{2}} \sigma_{ij} d\eta \; , \; M_{ij} = \int_{-\frac{h}{2}}^{\frac{h}{2}} \sigma_{ij} \eta \, d\eta \; \cdots \tag{7.66 a-b}$$

Note that they are defined over a cross-section with thickness h and unity width.

The Ilyushin yield condition proposed by Robinson [7.18] is stated as

$$Q_t + Q_m + \frac{|Q_{tm}|}{\sqrt{3}} \leq 1 \cdots \tag{7.67}$$

with

$$Q_t = n_{11}^2 + n_{22}^2 - n_{11}n_{22} + 3n_{12}^2$$

$$Q_m = m_{11}^2 + m_{22}^2 - m_{11}m_{22} + 3m_{12}^2 \qquad \cdots \qquad \text{(7.68 a-c)}$$

$$Q_{tm} = n_{11}m_{11} + n_{22}m_{22} - \frac{1}{2}n_{11}m_{22} - \frac{1}{2}m_{11}n_{22} + 3n_{12}m_{12}$$

It should be noted that in the above yield condition transverse shear stresses have been disregarded. Robinson has shown that Eq. (7.67) is a very good approximation to the exact criterion [7.18].

7.5.4. Return Mapping

When the Ilyushin yield condition of Eq. (7.67) is violated, the state of stress falls outside of the yield surface described by Eq. (7.67). Such stresses have to be "brought back" to the yield surface. This is known as returning to the yield surface or return mapping. In the present investigation, the return mapping described in Ref. [7.19] is adopted. Specifically, the backward-Euler scheme of Sec. 6.6.6 of [7.19] is employed. With reference to Fig. **7.1**, the state of stress at A is known. Adding an elastic stress increment to stress state A results in stress state B, which falls outside of the yield surface. The return mapping then determines the "plastic strain increment multiplier" $\Delta\lambda$:

$$\Delta\lambda = \frac{f_B}{a_B^T \tilde{D} a_B + H'} \cdots \qquad (7.69)$$

such that stresses are returned to stress state C which is much closer to the yield surface than stress state B,

$$\sigma_C = \sigma_B - \Delta\lambda \tilde{D} a_B \cdots \qquad (7.70)$$

In Eqs. (7.69) and (7.70), f_B is the value of yield function at B, $\tilde{D}$ the elastic matrix of Eq. (7.57), $H' = E\frac{E_T}{E-E_T}$ and a_B is a vector representing the outward normal to the yield surface at B.

Considering the Ilyushin yield condition of Eq. (7.67), the state of stress is represented by membrane forces and bending moments. That is, the σ in Eq. (7.70) becomes

$$\sigma \to \left[N_{11} \quad N_{22} \quad N_{12} \quad M_{11} \quad M_{22} \quad M_{12} \right]^{T} \cdot_{\dots} \tag{7.71}$$

At the same time, the elastic matrix $\tilde{D}$ becomes,

$$\tilde{D} \to \begin{bmatrix} D_3 & & \\ & \dfrac{h^3}{12} D_3 & \\ & & \end{bmatrix}_{\dots} \tag{7.72 a}$$

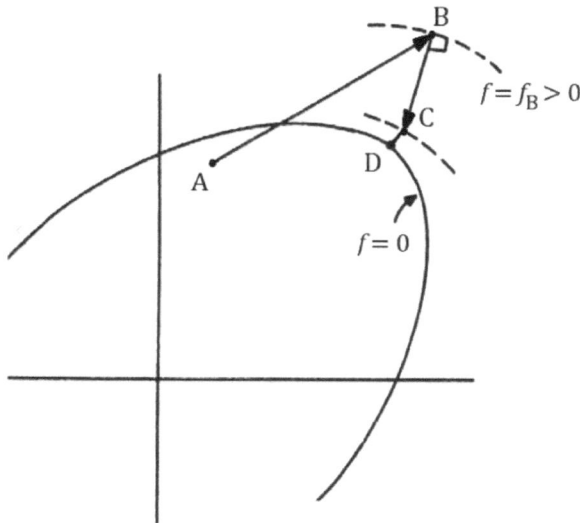

Figure 7.1: Backward-Euler scheme of return mapping.

where

$$D_3 = \begin{bmatrix} \dfrac{E}{1-v^2} & \dfrac{vE}{1-v^2} & \\ \dfrac{vE}{1-v^2} & \dfrac{E}{1-v^2} & \\ & & \dfrac{E}{2(1+v)} \end{bmatrix}_{\dots} \tag{7.72 b}$$

Note that D_3 is the standard 3×3 elastic matrix for plane stress, and $\frac{h^3}{12} D_3$ is the standard 3×3 elastic matrix for bending.

Then the yield function as applied to Eq. (7.69) and the normal vector are, omitting the subscript "*B*" [7.20]

$$f = Q_t + Q_m + \frac{1}{\sqrt{3}}|Q_{tm}| - 1 \; , \; a = \frac{1}{2h^4\sigma_e}\begin{Bmatrix} h^2 N_{,N} + \frac{2sh}{\sqrt{3}} M_{,M} \\ \frac{2sh}{\sqrt{3}} N_{,N} + 16 M_{,M} \end{Bmatrix} \ldots \qquad \text{(7.73 a-b)}$$

where

$$N_{,N} = \begin{Bmatrix} 2N_{11} - N_{22} \\ 2N_{22} - N_{11} \\ 6N_{12} \end{Bmatrix} \; , \; M_{,M} = \begin{Bmatrix} 2M_{11} - M_{22} \\ 2M_{22} - M_{11} \\ 6M_{12} \end{Bmatrix} \; ,$$

$$\ldots \qquad \text{(7.73 c-f)}$$

$$s = \frac{Q_{tm}}{|Q_{tm}|} \; , \; \sigma_e = \sqrt{h^2 Q_t + 16h^4 Q_m + \frac{4h^3}{\sqrt{3}}|Q_{tm}|}$$

and N_{ij} and M_{ij} are the membrane forces and bending moments defined in Eq. (7.66).

It should be mentioned that the above backward-Euler scheme can be applied again and again (in Fig. **7.1**, *C* would be the new *B*, and *D* would become the new *C*, and so on). Numerical experiments in the present investigation have found that, in a very large majority of cases in which the return mapping must be exercised, the backward-Euler scheme needs to be applied once. Cases requiring three or more applications of the scheme could not be recalled at the time of writing. In the present investigation, a state of stress is considered on the yield surface when abs(*f*)/σ_e < 10^{-9}, where abs(*f*) is the absolute value of the yield function at "*C*" and evaluated by Eq. (7.73a), and σ_e is the equivalent stress at "*C*" and determined by Eq. (7.73f).

7.6. CONFIGURATION AND STRESS UPDATING

This section is concerned with the updating of configuration and stress at every time step. The updating of mid-surface coordinates and directors is introduced in Sub-sec. 7.6.1 while the updating of stress is presented in Sub-sec. 7.6.2. For dynamic analyses, the linear velocities and accelerations, as well as the director angular velocity field and director angular acceleration field also need to be updated. Such updating will be included in Sec. 7.7.

7.6.1. Updating of Configuration

The updating of configuration of a shell consists of the updating of its mid-surface coordinates and its directors. Mid-surface coordinates are updated by adding mid-surface displacements to mid-surface coordinates of the reference configuration. Setting η to zero in Eqs. (7.13) and (7.15), one can write, for mid-surface coordinates of the reference configuration

$$\begin{Bmatrix} r \\ s \\ t \end{Bmatrix} = \sum_{i=1}^{3} \xi_i \begin{Bmatrix} r_{0i} \\ s_{0i} \\ 0 \end{Bmatrix} \cdots \qquad (7.74\ a)$$

and for incremental displacements

$$\begin{Bmatrix} \Delta u \\ \Delta v \\ \Delta w \end{Bmatrix} = \sum_{i=1}^{3} \xi_i \begin{Bmatrix} \Delta u_{0i} \\ \Delta v_{0i} \\ \Delta w_{0i} \end{Bmatrix} \cdots \qquad (7.74\ b)$$

Adding displacement increments to mid-surface coordinates of the reference configuration results in the following updating scheme for mid-surface coordinates

$$\begin{Bmatrix} r^{t+\Delta t} \\ s^{t+\Delta t} \\ t^{t+\Delta t} \end{Bmatrix} = \begin{Bmatrix} r \\ s \\ t \end{Bmatrix} + \begin{Bmatrix} \Delta u \\ \Delta v \\ \Delta w \end{Bmatrix} = \sum_{i=1}^{3} \xi_i \begin{Bmatrix} r_{0i} \\ s_{0i} \\ 0 \end{Bmatrix} + \sum_{i=1}^{3} \xi_i \begin{Bmatrix} \Delta u_{0i} \\ \Delta v_{0i} \\ \Delta w_{0i} \end{Bmatrix} \cdots \qquad (7.75)$$

Note that Eqs. (7.74) and (7.75) are written in terms of local coordinates. Similar construction for the global coordinates can be made. For brevity, those in terms of global coordinates will not be presented here.

The updating procedures for the directors follow those proposed in [7.3-7.4]. Assuming that the incremental rotation vector $\Delta\theta$ at a node has been solved, $V^{t+\Delta t}$, the new position of director V, is updated through the following equation

$$V^{t+\Delta t} = T_v V \; ... \tag{7.76}$$

with

$$T_v = \cos\left(|\Delta\theta|\right) I_3 + \frac{\sin\left(|\Delta\theta|\right)}{|\Delta\theta|} \widehat{\Delta\theta} \; ... \tag{7.77 a}$$

where $|\cdot|$ denotes the magnitude of the enclosed vector and $\widehat{\Delta\theta}$ is a skew-symmetric matrix constructed from $\Delta\theta$. That is,

$$\widehat{\Delta\theta} = \begin{bmatrix} & -\Delta\theta_t & \Delta\theta_s \\ \Delta\theta_t & & -\Delta\theta_r \\ -\Delta\theta_s & \Delta\theta_r & \end{bmatrix} \; ... \tag{7.77 b}$$

It should be mentioned that in the limiting case of small rotations when $\Delta\theta \to 0$, since $\cos\left(|\Delta\theta|\right) \to 1$ and $\dfrac{\sin\left(|\Delta\theta|\right)}{|\Delta\theta|} \to 1$, it can be seen that $T_v \to I_3$. As a result, the updating scheme for directors reduces to that for translational displacements.

Finally, as a part of configuration updating, the updating of mass density and thickness must be considered. Such an updating requires the calculation of "relative" deformation gradient which is defined as

$$F_t^{t+\Delta t} = \begin{vmatrix} 1 + \dfrac{\partial \Delta u}{\partial r} & \dfrac{\partial \Delta u}{\partial s} & \dfrac{\partial \Delta u}{\partial t} \\ \dfrac{\partial \Delta v}{\partial r} & 1 + \dfrac{\partial \Delta v}{\partial s} & \dfrac{\partial \Delta v}{\partial t} \\ \dfrac{\partial \Delta w}{\partial r} & \dfrac{\partial \Delta w}{\partial s} & 1 + \dfrac{\partial \Delta w}{\partial t} \end{vmatrix} \; ... \tag{7.78}$$

where displacement increments $\Delta u, \Delta v, \Delta w$ are as given in Eq. (7.15). Equation (7.81) expresses the deformations of the body occupying configuration $C^{t+\Delta t}$ with respect to the reference configuration C^t. Subsequently, mass density, thickness and area of the mid-surface can be updated:

$$\rho^{t+\Delta t} = \frac{\rho^t}{\det\left(F_t^{t+\Delta t}\right)} \; , \; h^{t+\Delta t} = h^t \det\left(F_t^{t+\Delta t}\right)\frac{A^t}{A^{t+\Delta t}} \; , \; A^{t+\Delta t} = \frac{r^{t+\Delta t} s^{t+\Delta t}}{2} \cdots \qquad (7.79\text{ a-c})$$

with det($\cdot$) denoting the "determinant of". Quantities associated with time "t" are superscripted in Eq. (7.79) for clarity.

7.6.2. Updating of Stress

As Eqs. (7.11) and (7.12) indicated, after solving the nodal displacement increments, strain and stress increments can be recovered. This stress increment is defined with respect to the reference configuration C^t, and can therefore be added to σ, the Cauchy stress at time "t" since they are referred to the same reference state. The sum of σ and ΔS becomes $S^{t+\Delta t}$. That is, $S^{t+\Delta t} = \sigma + \Delta S$, which is the second Piola-Kirchhoff stress of deformation state $C^{t+\Delta t}$ measured with respect to the reference configuration C^t. The transformation of the second Piola-Kirchhoff stress $S^{t+\Delta t}$ to the Cauchy stress $\sigma^{t+\Delta t}$ is [7.21]

$$\sigma^{t+\Delta t} = \frac{F_t^{t+\Delta t} S^{t+\Delta t} \left(F_t^{t+\Delta t}\right)^T}{\det\left(F_t^{t+\Delta t}\right)} \cdots \qquad (7.80)$$

where $F_t^{t+\Delta t}$ has been defined by Eq. (7.78).

7.7. NUMERICAL ALGORITHMS

While the focus of this and the next two chapters is on the nonlinear dynamic analyses, nonlinear static analyses are deemed necessary from time to time for the understanding or interpretation of the dynamic behaviour of the structure under consideration. This section will cover briefly the numerical algorithms employed in the present investigation. Sub-sec. 7.7.1 deals with algorithms for nonlinear static analyses. Algorithms for nonlinear dynamic analyses will be presented in Sub-sec. 7.7.2.

7.7.1. Algorithms for Nonlinear Static Analysis

Algorithms for nonlinear static analysis typically include [7.19, 7.21], for example, the full Newton-Raphson method, the modified Newton-Raphson method, the initial stress method and the Broyden-Fletcher-Goldfarb-Shanno (BFGS) method. To be able to deal with snap-through buckling, the so-called arc-length methods are employed. In the present investigation, the Riks-Wempner arc-length method is adopted. A brief outline of the method is given below. For further details, readers are to refer to [7.19, 7.22-7.23], for example.

With reference to Fig. **7.2**, assuming that the displacements and the state of stress at "i" are known and converged. The corresponding global stiffness matrix is K_i. If F_e is the external load vector at full load level, displacement increment vector ΔQ_i due to the prescribed load increment can be determined by

$$K_i\left(\Delta Q_i\right)=\left(\Delta \lambda_i\right)F_e\,...$$ (7.81)

Where $\Delta \lambda_i$ is a scalar multiplier such that $\Delta \lambda_i F_e$ gives the increment in the external load. Typically, $\Delta \lambda_i = 0.01$. Note that the $\Delta \lambda_i$ or $\Delta \lambda_k$ in this sub-section should not be confused with the "plastic strain increment multiplier" of Sub-sec. 7.5.4.

Subsequently, the applied load level and displacements are updated to "k":

$$\lambda_k = \lambda_i + \Delta \lambda_i \ , \quad Q_k = Q_i + \Delta Q_i\,...$$ (7.82 a-b)

However, the displacements and state of stress at "k" are in general not converged. The Riks-Wempner arc-length method then seeks to find displacements along the direction n_i (which is normal to the tangent vector t_i) that possesses better convergence. The condition of $t_i \cdot n_i = 0$ (where "$\cdot$" indicates dot product) leads to

$$\Delta Q_k = \Delta Q_k^{II} - \Delta \lambda_k \Delta Q_k^{I}\,...$$ (7.83 a)

where ΔQ^I and ΔQ^{II} are determined by

$$K_k\left(\Delta Q^I\right) = F_e \ , \quad K_k\left(\Delta Q^{II}\right) = \lambda_k F_e - F_1 \ldots \tag{7.83 b-c}$$

In Eqs. (7.83b-c), K_k is the global stiffness matrix at "k", and F_1 is the global internal force vector at "k". The scalar λ_k indicates the load level at "k". Finally, the scalar $\Delta \lambda_k$ in Eq. (7.83a) is determined by

$$\Delta \lambda_k = \frac{\left(\Delta Q_i\right)^T \left(\Delta Q_k^{II}\right)}{\left(\Delta Q_i\right)^T \left(\Delta Q_k^I\right) + \Delta \lambda_i} \ldots \tag{7.84}$$

With $\Delta \lambda_k$ now known, the applied load level and displacements are updated to "$k+1$"

$$\lambda_{k+1} = \lambda_k - \Delta \lambda_k \ , \quad Q_{k+1} = Q_k + \Delta Q_k \ldots \tag{7.85 a-b}$$

The convergence of displacements and state of stress at "$k+1$" is checked against some chosen convergence criteria. If such criteria are not met, the above scheme is repeated by resetting "$k+1$" to "k", $\Delta Q_i + \Delta Q_k \to \Delta Q_i$, and $\Delta \lambda_i + \Delta \lambda_k \to \Delta \lambda_i$. The commonly adopted convergence criteria include the normalized force-based and displacement-based convergence criteria suggested by Crisfield [7.19], with the β in Eqs. (9.48) and (9.49) of [7.19] being set to 0.001 to 0.01. In the present investigation, it has been found that, only the first few load increments require repeating in the above scheme. Typically, it requires no more than 5 iterations. For the remaining load increments, convergence is achieved at reaching "$k+1$" the first time. In addition, it has been found that applying the above scheme only once per load increment yields very little difference. For computational efficiency, the above scheme is applied once per load increment.

It should be noted that the displacement updating scheme as indicated by Eqs. (7.82b) and (7.85b), strictly speaking, is only applicable to translational DOF. Directors, on the other hand, should be updated through Eq. (7.76). This is particularly true for problems involving finite rotation.

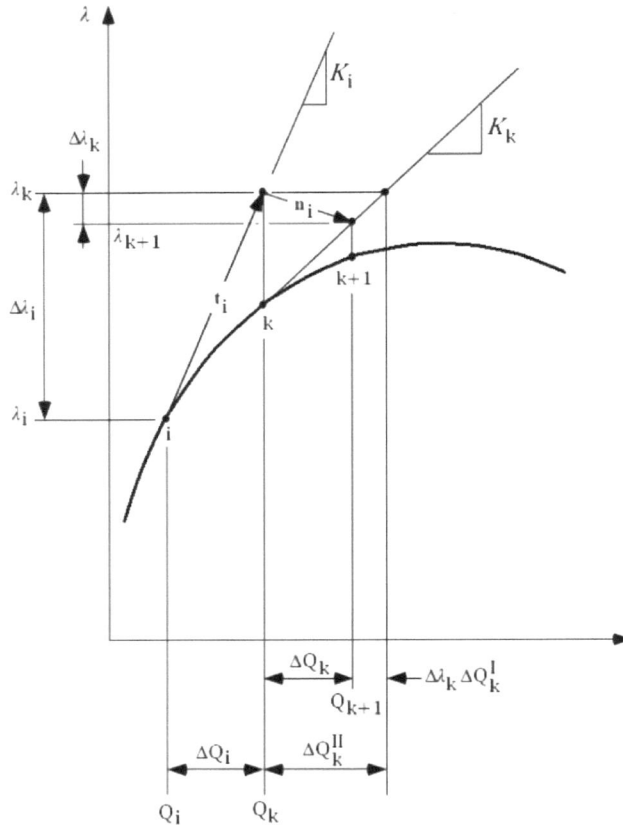

Figure 7.2: Riks-Wempner arc-length method.

7.7.2. Algorithms for Nonlinear Dynamic Analysis

Amongst direct time integration schemes, the Newmark family of algorithms and their variants are perhaps the best known for their relative simplicity of implementation and efficiency. The details of their implementations can be found in [7.21, 7.24], for example. These algorithms, in their original forms, are however limited to small rotation, small director angular velocity and small director angular acceleration. Under the term "conserving algorithms", Ref. [7.9] proposed an extension of the Newmark-α family of algorithms to large rotation, large director angular velocity and large director angular acceleration. Of course, the only member of the family of algorithms that conserves linear as well angular moment a is that known as the mid-point rule described by parameter setting of: α

= 1/2, $2\beta/\gamma = 1$ [7.9] where α, γ and β are parameters of the Newmark-α family of direct integration algorithms.

The scheme to update the linear velocities and linear accelerations is [7.9],

$$
\dot{U}^{t+\Delta t} = \frac{\gamma}{\alpha\beta\Delta t}\Delta U - \left[\dot{U}^t + \left(\frac{\gamma}{\beta} - 2\right)\left(\dot{U}^t + \frac{\Delta t}{2}\ddot{U}^t\right)\right]
$$
$$
\ddot{U}^{t+\Delta t} = \frac{1}{\gamma\Delta t}\dot{U}^{t+\Delta t} - \left[\frac{1}{\gamma\Delta t}\dot{U}^t + \left(\frac{1}{\gamma} - 1\right)\ddot{U}^t\right]
$$
... (7.86 a)

where $\Delta U, U, \dot{U}$ and $\ddot{U}$ are the global incremental linear displacement vector, global linear displacement vector, global linear velocity vector and global linear acceleration vector, respectively. Note that $\Delta U, U, \dot{U}$ and $\ddot{U}$ are extracted from $\Delta Q, Q, \dot{Q}$ and $\ddot{Q}$ which are the global incremental displacement vector, global displacement vector, global velocity vector and global acceleration vector, respectively.

On the other hand, the director angular velocity field $\omega^t = V^t \times \dot{V}^t$ and director angular acceleration field $\dot{\omega}^t$ are to be updated, by the following scheme:

$$
\omega^{t+\Delta t} = \frac{\gamma}{\alpha\beta\Delta t}\Delta\theta - T_v\left[\omega^t + \left(\frac{\gamma}{\beta} - 2\right)\left(\omega^t + \frac{\Delta t}{2}\dot{\omega}^t\right)\right] ,
$$
$$
\dot{\omega}^{t+\Delta t} = \frac{1}{\gamma\Delta t}\omega^{t+\Delta t} - T_v\left[\frac{1}{\gamma\Delta t}\omega^t + \left(\frac{1}{\gamma} - 1\right)\dot{\omega}^t\right] .
$$
... (7.86 b)

Note that the matrix T_v is constructed from $\Delta\theta$ according to Eq. (7.77). Since $\Delta\theta$ is the incremental rotation vector at any given node, Eq. (7.86b) needs to be applied node by node. Finally, the Δt in Eqs. (7.86a-b) is the time step size.

Two comments are in order. Firstly, Eq. (7.86) is not applicable to the central difference method owing to the fact that $\beta = 0$ for the central difference method. Secondly, since the present investigation is limited to cases in which angular velocities and accelerations are small (but not necessarily small rotations), it has been found through numerical experiments that using the T_v matrix as defined by

Eq. (7.77) and setting it to a 3×3 identity matrix result in little difference. However, the latter enables the updating director angular velocities and director angular accelerations at the global level. That is, Eq. (7.86a) is applied with $\Delta U, U, \dot{U}, \ddot{U}$ replaced by $\Delta Q, Q, \dot{Q}$ and $\ddot{Q}$.

Another implicit direct time integration scheme that has found success in the present investigation is the backward-Euler scheme by Liu *et al.* [7.25]. Considering the case of zero damping, the scheme determines the effective stiffness matrix and effective load vector as follows:

$$\tilde{K}^{t+\Delta t} = M^t + \frac{1}{2}(\Delta t)^2 K^t$$

$$\tilde{F}^{t+\Delta t} = M^t \left(\Delta t \dot{Q}^t\right) + \frac{1}{2}(\Delta t)^2 \left(F_e^{t+\Delta t} - F_1\right) \qquad (7.87)$$

Next, the global incremental displacement vector ΔQ is solved

$$\tilde{K}^{t+\Delta t} \left(\Delta Q\right) = \tilde{F}^{t+\Delta t} \, ... \qquad (7.88)$$

and displacement, velocity and acceleration are updated

$$Q^{t+\Delta t} = Q^t + \Delta Q$$

$$\dot{Q}^{t+\Delta t} = \begin{cases} \dfrac{3}{\Delta t}\left(Q^{t+\Delta t} - Q^t\right) - \dfrac{1}{2}\dot{Q}^t + \dfrac{\Delta t}{4}\ddot{Q}^t & t = 0 \\[2mm] \dfrac{3}{2\Delta t}Q^{t+\Delta t} - \dfrac{2}{\Delta t}Q^t + \dfrac{1}{2\Delta t}Q^{t-\Delta t} & t > 0 \end{cases} \, ... \qquad (7.89)$$

$$\ddot{Q}^{t+\Delta t} = \frac{2}{(\Delta t)^2}\left[Q^{t+\Delta t} - Q^t - \Delta t \dot{Q}^t\right]$$

In closing, the trapezoidal rule and the backward-Euler scheme of Liu, *et al.* are employed for the computations in the following two chapters.

DISCLOSURE

Part of the information included in this chapter has been previously published in *Computers & Structures, Volume 54, Issue 6, March, 1995, Pages 1031-1056.*

REFERENCES

[7.1]   M.L. Liu and C.W.S. To, "Hybrid strain based three-node flat triangular shell elements, Part I: Nonlinear theory and incremental formulations", *Comp. Struct.* Vol. 54, pp. 1031-1056, March 1995.

[7.2]   A.F. Saleeb, T.Y. Chang, W. Graf, and S. Yingyeunyong, "A hybrid/mixed model for non-linear shell analysis and its applications to large rotation problems", *Int. J. Numer. Methods Eng.*, vol. 29, pp. 407-446, February 1990.

[7.3]   J.C. Simo, and D.D. Fox, "On a stress resultant geometrically exact shell model. Part I: Formulation and optimal parametrization", *Comp. Methods Appl. Mech. Eng.*, vol. 72, pp. 267-304, March 1989.

[7.4]   J.C. Simo, D.D. Fox, and M.S. Rifai, "On a stress resultant geometrically exact shell model. Part II: The linear theory; Computational aspects", *Comp. Methods Appl. Mech. Eng.*, vol. 73, pp. 53-92, April 1989.

[7.5]   D.J. Allman, "A compatible triangular element including vertex rotations for plane elasticity analysis", *Comp. Struct.*, vol. 19, pp. 1-8, January1984.

[7.6]   A. Tessler, and T.J.R. Hughes, "A three-node Mindlin plate element with improved transverse shear", *Comp. Methods Appl. Mech. Eng.*, vol. 50, pp. 71-101, July 1985.

[7.7]   C.W.S.To, and M.L. Liu, "Hybrid strain based three-node flat triangular shell elements", *Finite Elements in Analysis and Design*, vol. 17, pp. 169-203, October 1994.

[7.8]   M.L. Liu, "Response statistics of shell structures with geometrical and material nonlinearities",Ph.D. thesis, The University of Western Ontario, London, ON, 1993.

[7.9]   J.C. Simo, M.S. Rifai, and D.D. Fox, "On a stress resultant geometrically exact shell model. Part VI: Conserving algorithms for nonlinear dynamics", *J. Numer. Methods Eng.*, vol. 34, pp. 117-164, March 1992.

[7.10] T.J.R. Hughes, and W.K. Liu, "Nonlinear finite element analysis of shells, Part I: Three dimensional shells", *Comp. Methods Appl. Mech. Eng.*, vol. 26, pp. 331-362, June 1981.

[7.11] J.W. Hutchinson, "Finite strain analysis of elasto-plastic solids and structures", in *Numerical Solution of Nonlinear Structural Problems*, 1975, pp.17-29.

[7.12] J.C. Nagtegaal, and J.E. de Jong, "Some computational aspects of elastic-plastic large strain analysis", *J. Numer. Methods Eng.*, vol. 17, pp. 15-41, January 1981.

[7.13] A. Needleman, "A numerical study of necking in circular cylindrical bars", *J. Mech. Phys. Solids*, vol. 20, pp. 111-127, May 1972.

[7.14] D.R.J. Owen, and E. Hinton, *Finite Elements in Plasticity: Theory and Practice*, Swansea: Pineridge Press, 1980.

[7.15] J.C. Simo, and J.G. Kennedy, "On a stress resultant geometrically exact shell model. Part V: Nonlinear plasticity; Formulation and integration algorithms", *Comp. Methods Appl. Mech. Eng.*, vol. 96, pp. 133-171, April 1992.

[7.16] A.A. Ilyushin, *Plasticity* (in Russian). Moscow: Gostekhizdat, 1948.

[7.17] G.S. Shapiro, "On yield surfaces for ideally plastic shells", in *Problems of Continuum Mechanics*, 1961, pp.414-418.

[7.18] M. Robinson, "A comparison of yield surfaces for thin shells", *Int. J. Mech. Sci.*, vol. 13, pp. 345-354, April 1971.

[7.19] M.A. Crisfield, *Non-linear Finite Element Analysis of Solids and Structures, Vol. I: Essentials*. West Sussex: Wiley, 1991.

[7.20] M.A. Crisfield, *Non-linear Finite Element Analysis of Solids and Structures, Vol. II: Advanced Topics*. West Sussex: Wiley, 1997.

[7.21] K.J. Bathe, *Finite Element Procedures in Engineering Analysis*. New York: Prentice-Hall, 1982.

[7.22] E. Riks, "An incremental approach to the solution of snapping and buckling problems", *Int. J. Solids Struct.*, vol.15, pp. 529-551, July 1979.

[7.23] G.A. Wempner, "Discrete approximations related to nonlinear theories of solids", *Int. J. Solids Struct.*, vol. 7, pp. 1581-1599, November 1971.

[7.24] O.C. Zienkiewicz, and R.L. Taylor, *The Finite Element Method, Volume 1: The Basis* (5th Edn). Oxford: Butterworth-Heinemann, 2000.

[7.25] T. Liu, C. Zhao, Q. Li, and L. Zhang, "An efficient backward Euler time-integration method for nonlinear dynamic analysis of structures", *Comp. Struct.* vol. 106/107, pp. 20-28, September 2012.

Send Orders for Reprints to reprints@benthamscience.net
Vibration and Nonlinear Dynamics of Plates and Shells, 2014, 149-165   **149**

CHAPTER 8

Nonlinear Dynamics of Flat-Surface Structures

Abstract: This chapter deals with the nonlinear dynamic responses of structures with flat mid-surfaces, such as plates and boxes by employing the mixed formulation based three-node flat triangular nonlinear shell elements presented in Chapter 7. Geometrical nonlinearity due to large deformation, material nonlinearity due to elastic-plastic material behaviour, and various loading situation including non-conservative loads, will be investigated.

Keywords: Dynamics, nonlinearities, geometrical, material, flat-surface.

8.1. INTRODUCTION

In this and the next chapters, nonlinear responses of various structures, under different loading scenarios, having geometrical as well as material nonlinearities will be examined. Results will be presented in two chapters. This chapter deals with structures with flat mid-surface or mid-surfaces, such as plates and boxes. Structures with single curvature or double curvatures will be investigated in the next chapter. For brevity, plots of meshes will be omitted unless deemed necessary. Similarly, detailed boundary conditions will not be listed. Readers may refer to Chapters 3 through 6 for such information.

8.2. CANTILEVER PLATE SUBJECTED TO UNIFORM PRESSURE

This plate has a length of 0.4 m, a width of 0.2 m and a plate thickness of 1 mm, see Fig. **8.1(a)**. Its material properties are: Young's modulus E = 70 GPa, Poisson's ratio v = 0.33 and mass density ρ_0 = 2700 kg/m^3. For finite element representation, a 8×4D mesh is employed to discretize the entire plate. The mesh has 8 divisions along the length of the plate and 4 divisions along its width. The plate is under a time-dependent load that is uniformly distributed over the entire plate. Owing to the nature of the problem, the two in-plane displacements, U and V, and the rotation about Z-axis, Θ_Z, are constrained to zero. For nodes located on the clamped edge, W, Θ_X and Θ_Y are also constrained to zero, resulting in an effective DOF of 216. Note that capital alphabets indicate displacements or rotations in the global coordinates, unless stated otherwise.

Meilan Liu and Cho W. S. To

A linear dynamic analysis is first performed with the maximum value of the uniform load being set to 500 Pa. The time history of the load is given in Fig. **8.1(b)**. The time step size for time integration is $\Delta t = 0.004$ s. The computed tip (vertical) displacement is shown in Fig. **8.2**. This displacement is found to be in very good agreement with results reported by Almeida and Awruch [8.1], for example. The latter reference showed an amplitude between 0.15 m and 0.155 m, and an average period of 0.185 second. It is noted that the time step size used in [8.1] is also $\Delta t = 0.004$ s.

Nonlinear dynamic analyses are subsequently conducted. Now the uniform load is linearly increased to 1000 Pa over the first 0.2 second, and remains at 1000 Pa afterward, see Fig. **8.1(c)**. Computation is performed with a reduced time step size of $\Delta t = 0.002$ s, treating the uniform pressure as conservative load and non-conservative load, respectively. Fig. **8.3** presents the nonlinear dynamic responses. It is seen that non-conservative load generally yields smaller amplitude of vibration. As time goes on, the plate will "settle onto" a different position from that taken under conservative load. However, it is interesting to observe the rather abrupt change, at the time instant of about 1 second, in the oscillation pattern of tip rotation, and horizontal displacement to a less extent. Further investigation is suggested to understand the nature and cause of such abrupt changes.

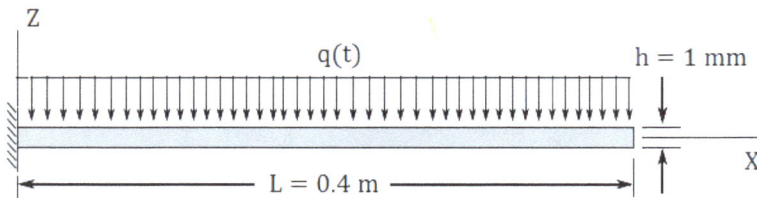

Figure 8.1(a): Geometry of cantilever plate.

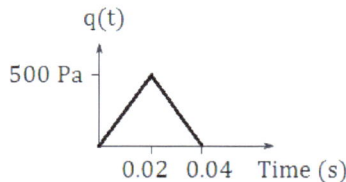

Figure 8.1(b): Time history of load for linear analysis.

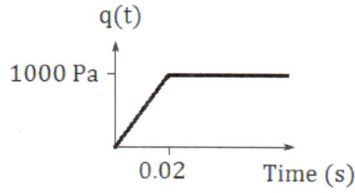

Figure 8.1(c): Time history of load for nonlinear analysis.

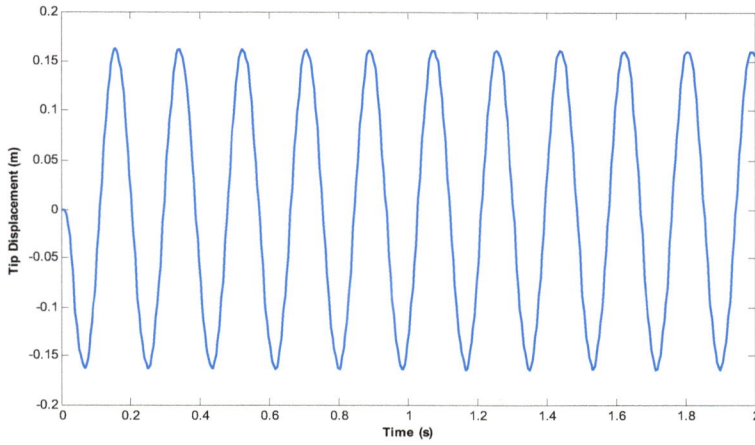

Figure 8.2: Linear responses of cantilever plate.

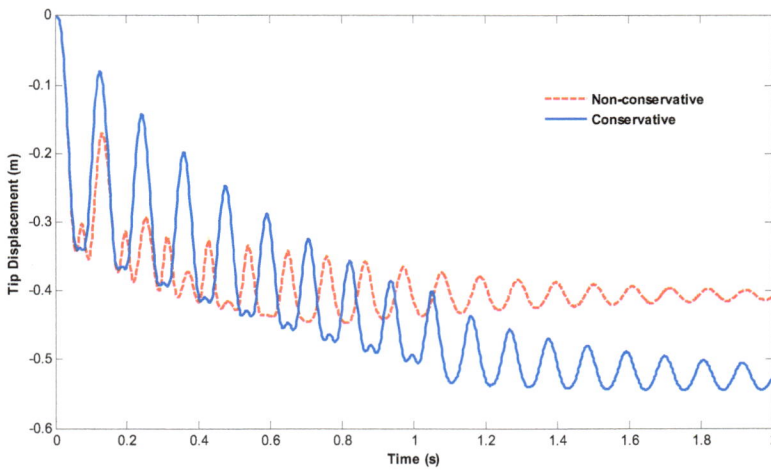

Figure 8.3(a): Nonlinear responses of cantilever plate (vertical displacement at tip).

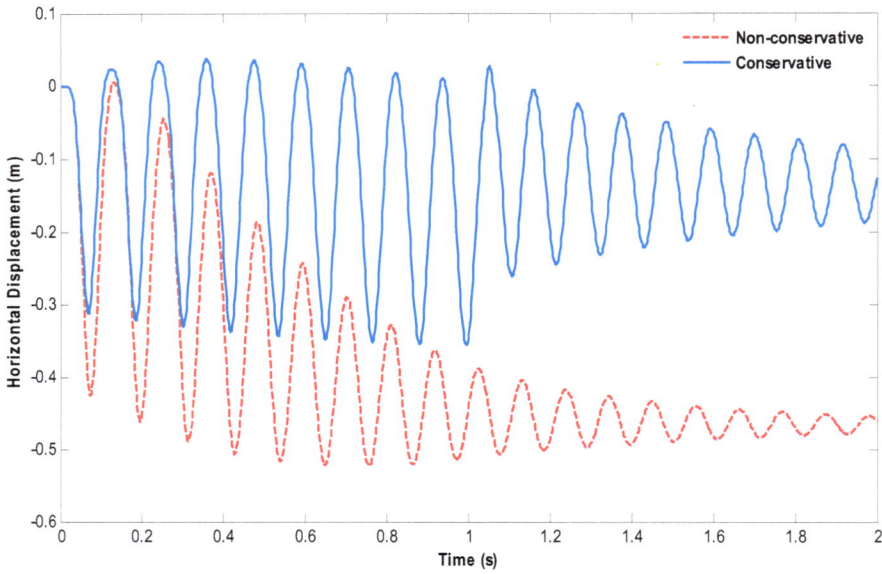

Figure 8.3(b): Nonlinear responses of cantilever plate (horizontal displacement at tip).

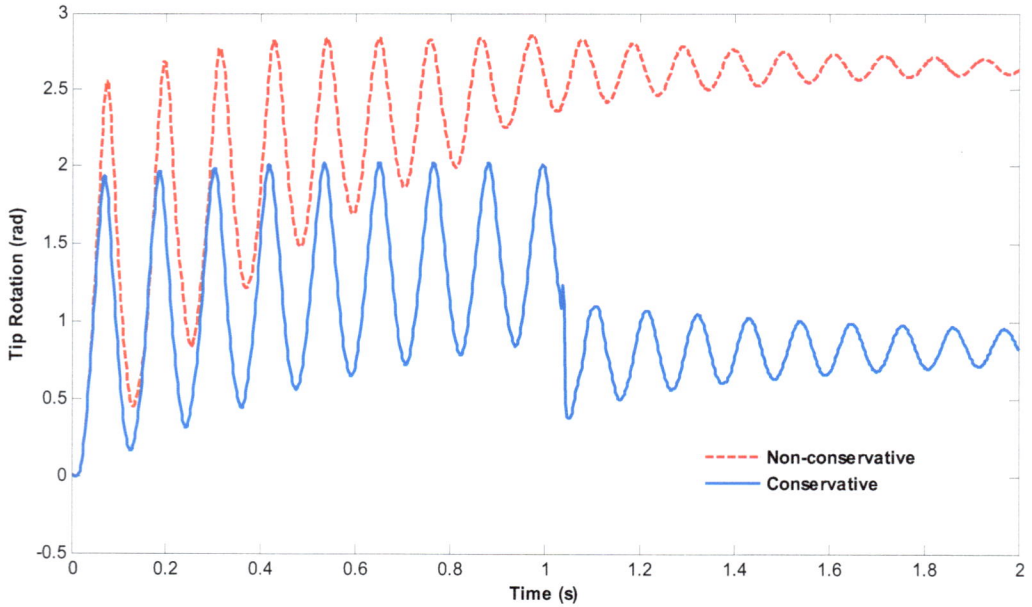

Figure 8.3(c): Nonlinear responses of cantilever plate (rotation Θ_Y at tip).

8.3. CLAMPED CIRCULAR PLATE SUBJECTED TO UNIFORM PRESSURE

The clamped circular plate (Fig. **8.4**) has a radius of $R = 2.54$ m (100 in) and a thickness of $h = 508$ mm (20 in). Therefore it is a moderately thick plate. The plate is subjected to a suddenly applied pressure of $q_0 = 689.5$ Pa (0.1 psi) (Fig. **8.5**) that is distributed uniformly over the entire plate. The material of the plate is elastic-perfectly plastic with the following properties: Young's modulus $E = 689.5$ kPa (100 psi), Poisson's ratio $v = 0.3$, yield strength $\sigma_Y = 10.343$ kPa (1.5 psi), and mass density $\rho_0 = 106.85 \times 10^6$ kg/m^3 (10 lb·s^2/in^4). These values may not be representative of engineering materials but were used by Huang in [8.2]. They are adopted for easy comparison. Computed elastic and elasto-plastic responses are given in Fig. **8.6** in which results from [8.2] are also included for comparison. The present formulation seems to predict higher peak values of central displacement and at earlier time instants. The elasto-plastic response is closer to those by [8.2] than the elastic counterpart is.

It should be mentioned that only a quarter of the circular plate is discretized, using type F mesh. Similar to the cantilever plate investigated in the previous section, the two in-plane displacements, U and V, and the rotation about Z-axis, Θ_Z, are constrained to zero. Nodes on the clamped edge have their W, Θ_X and Θ_Y constrained to zero as well. In addition, symmetry boundary conditions are applied to the two straight edges. As a result, the mesh has 496 effective DOF. The time step size is $\Delta t = 3$ s, as opposed to the much smaller $\Delta t = 0.6$ s used by [8.2]. Computation shows that using the smaller Δt yields minimal difference. That is, the present formulation allows for larger time step size, making it effective as well as accurate.

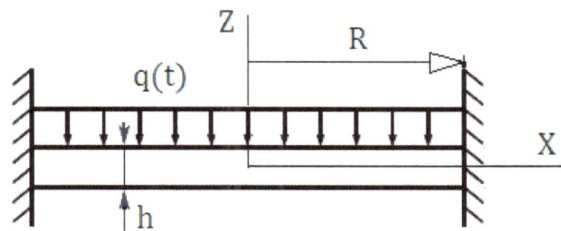

Figure 8.4: Clamped circular plate subjected to uniform pressure.

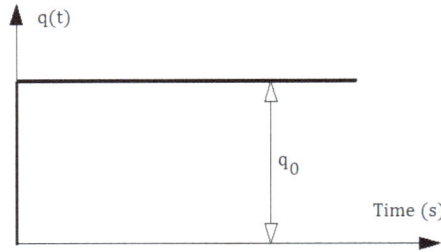

Figure 8.5: Suddenly applied load.

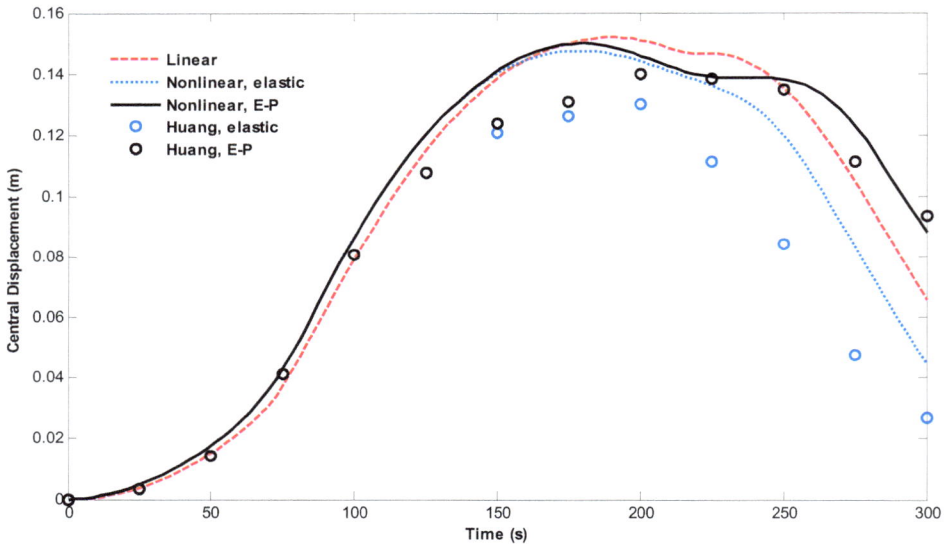

Figure 8.6: Time history of central displacement of clamped circular plate.

8.4. CLAMPED SQUARE PLATE SUBJECTED TO UNIFORM PRESSURE

The square plate under consideration here (Fig. **8.7**) has geometrical dimensions of $254 \times 254 \times 6.65$ mm^3 ($10 \times 10 \times 0.262$ in^3). It is therefore a thin plate. A 5×5D mesh is used to discretize a quarter of the plate. Again, U, V and Θ_Z are constrained to zero. Nodes on the clamped edges have their W, Θ_X and Θ_Y constrained to zero as well. Appropriate symmetry boundary conditions are then applied to the two remaining edges, resulting in an effective DOF of 140. The material is elastic-plastic with linear isotropic hardening. Properties are, $E = 207$

GPa (30×10^6 psi), $v = 0.3$, tangent modulus $E_T = 2.07$ GPa (3×10^5 psi), $\sigma_Y = 253$ MPa (36.7×10^3 psi), and $\rho_0 = 7800$ kg/m^3 (0.000733 lb·s^2/in^4).

The plate is subjected to a suddenly applied pressure of $q_0 = 1.17$ MPa (170 psi). Linear, nonlinear elastic and nonlinear elasto-plastic analyses are performed using a time step size of $\Delta t = 10^{-5}$ s. This is 10 times as large as that used by Daye and Toridis in [8.3]. Fig. **8.8** shows the computed time histories of the central displacement and compares with the elasto-plastic response of [8.3]. It should be noted that [8.3] did not list the q_0 value explicitly. But very good agreement is found at $q_0 = 1.17$ MPa (170 psi), as is seen in Fig. **8.8**.

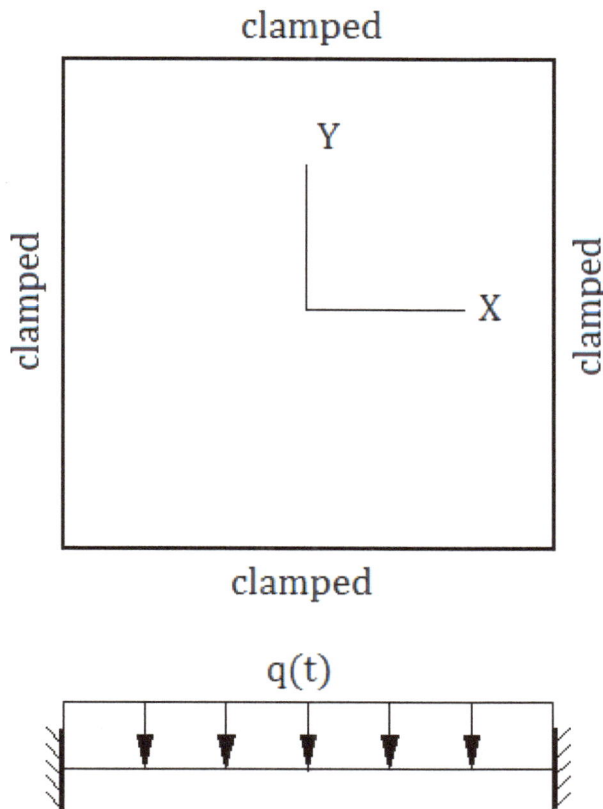

Figure 8.7: Clamped square plate subjected to uniform pressure.

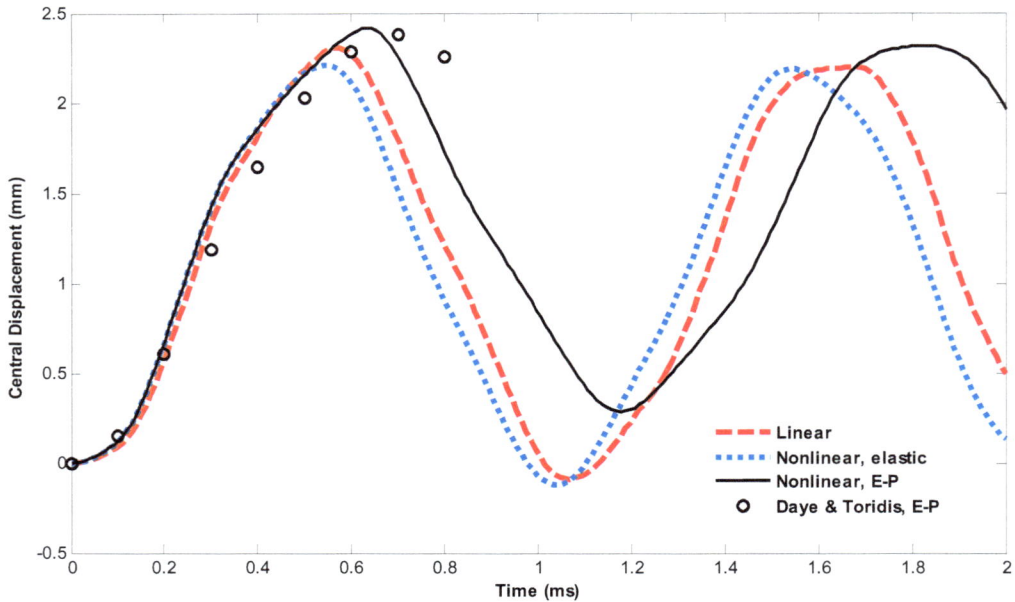

Figure 8.8: Central displacements of clamped square plate.

8.5. SIMPLY SUPPORTED RECTANGULAR PLATE SUBJECTED TO A CENTER LOAD

The plate is shown in Fig. **8.9**. Its geometrical dimensions are, a = 1016 mm (40 in), b = 1524 mm (60 in) and thickness h = 25.4 mm (1 in). The material properties are, Young's modulus E = 207 MPa (3×10^4 psi), Poisson's ratio ν = 0.25, yield stress σ_Y = 207×10^3 N/m² (30 psi) and mass per unit volume ρ_0 = 3210 kg/m³ (0.0003 lb-sec²/in⁴).

Owing to symmetry of geometry, boundary condition and loading, one quarter of the plate is discretized by a 8×12D mesh. After applying constraints, the effective DOF is 1113. The time step size is chosen to be Δt = 0.002 s as was in reference [8.4].

Two loading cases are considered. The first load case, Fig. **8.10(a)**, represents a step load that is increased from zero to its full value P_0 over a small time span t_0 = 0.006 s. Nonlinear elastic and elasto-plastic analyses are first performed with the center load being set to $P_0 = 178$ N (40 lb). Next, the center load is given larger values of 223 N, 267 N and 312 N (50 lb, 60 lb and 70 lb), and nonlinear elasto-plastic analyses are repeated. Figs. **8.11** and **8.12** show the computed central displacements of the plate. It can be seen that the plate experiences stiffness hardening as the center load gets larger and larger in magnitude. It should be noted that this rectangular plate was investigated in [8.4] for elasto-plastic response with $P_0 = 356$ N (80 lb). In the present computation, it is found, however, that the latter load would cause the plate to be completely plastic, resulting in singular matrices. It is thought that the discrepancy is attributed to the use of different return maps.

The second load case, as shown in Fig. **8.10(b)**, is to model an impulse. The time period t_0 remains 0.006 s and P_0 is 356 N (80 lb). Results of the nonlinear elastic and elasto-plastic analyses are given in Fig. **8.13**. They are found to be in agreement with those presented in [8.4].

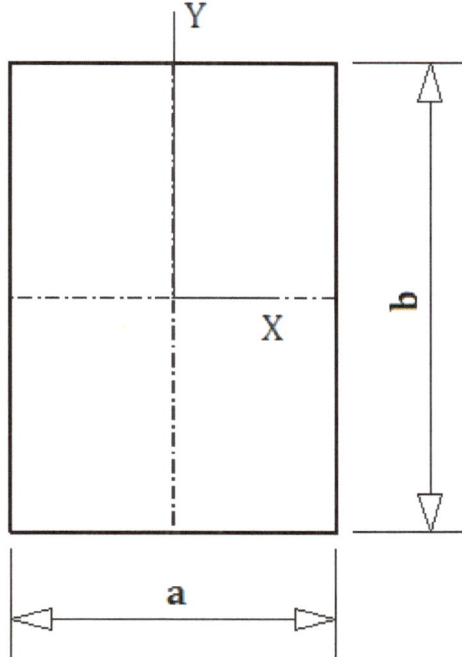

Figure 8.9: Simply supported rectangular plate.

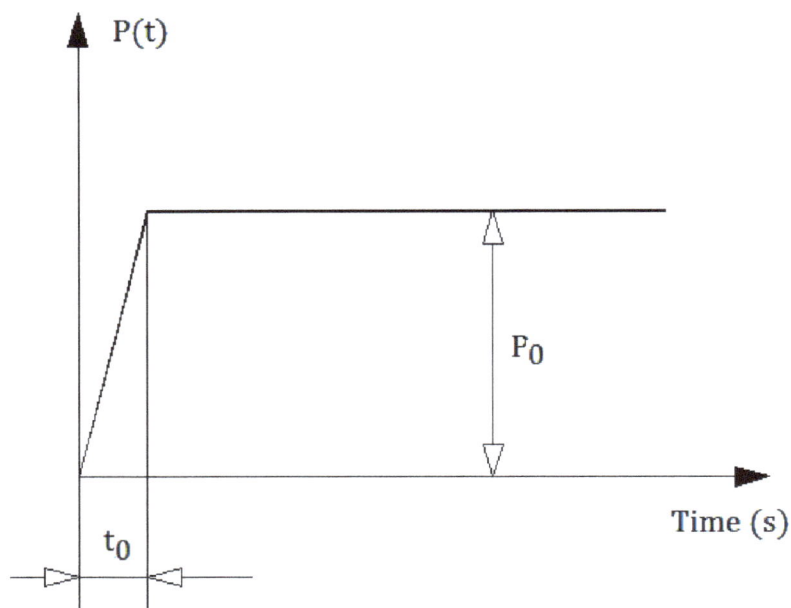

Figure 8.10(a): Time history of center load.

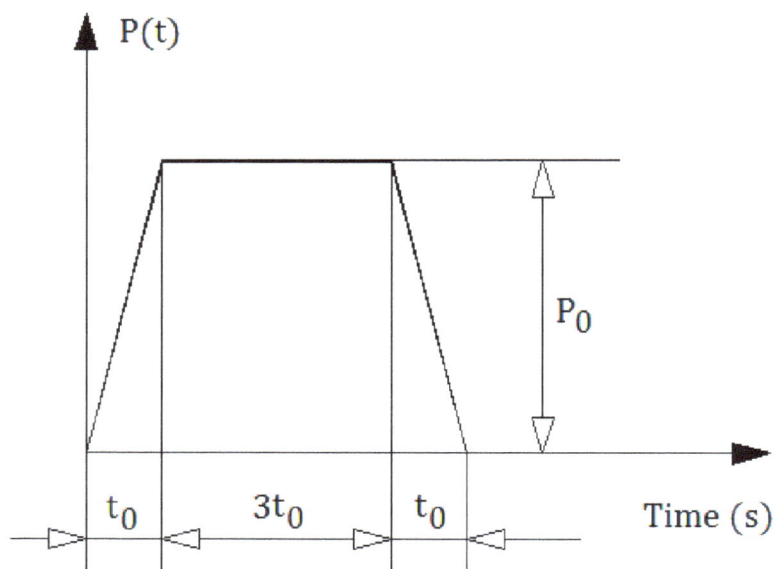

Figure 8.10(b): Time history of center load.

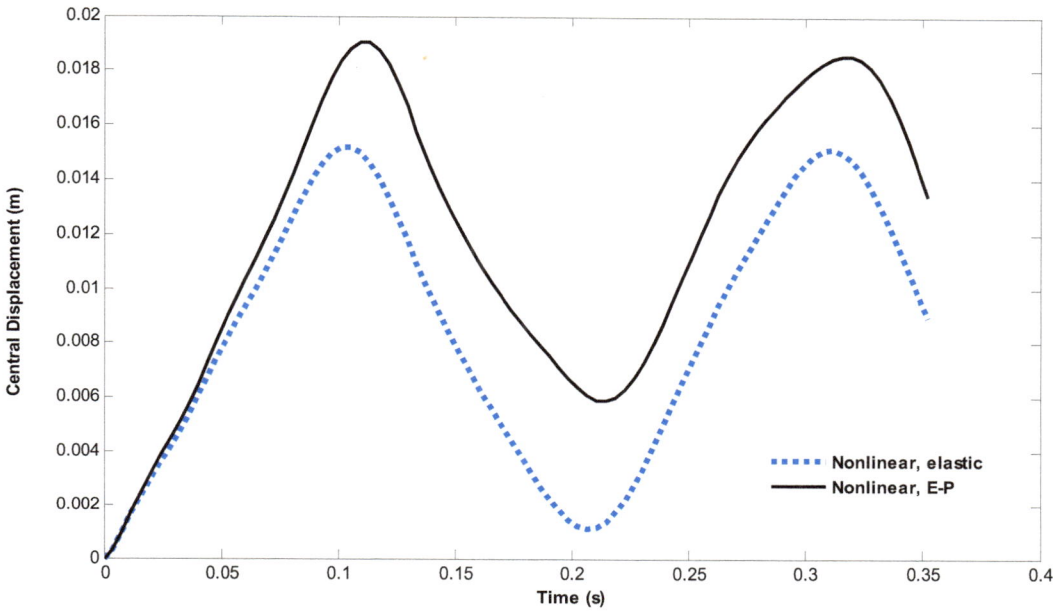

Figure 8.11: Central displacements of simply supported rectangular plate under step load ($P_0 = 178$ N).

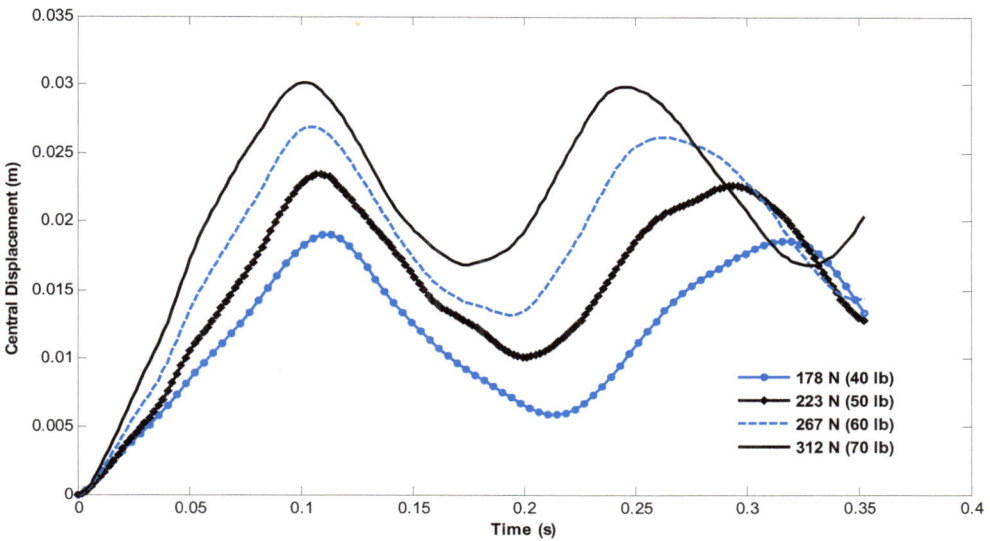

Figure 8.12: Elasto-plastic responses of simply supported rectangular plate under step loads.

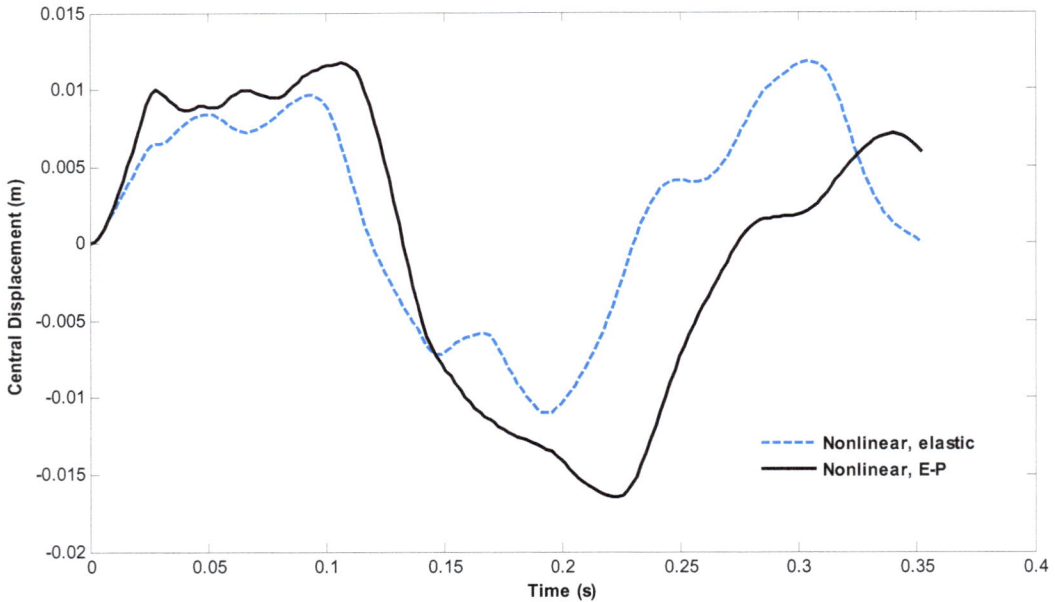

Figure 8.13: Elasto-plastic responses of simply supported rectangular plate under impulse-type of center load (P_0 = 356 N).

8.6. TUBE SUBJECTED TO INTERNAL AND EXTERNAL PRESSURE

As an example of nonlinear dynamics of structures consisting of flat mid-surfaces, a single-cell tube is considered. The tube, which is shown in Fig. **8.14**, has a cubic shape and a uniform thickness. The following material properties are used in computation: Young's modulus E = 205 GPa, Poisson's ratio v = 0.3, tangent stiffness E_T = 1.025 GPa, yield stress σ_Y = 210 MPa and mass per unit volume ρ_0 = 7900 kg/m^3. Due to symmetry, only one-eighth of the tube, the *ABCDEF* portion in Fig. **8.14**, is discretized, and the appropriate symmetry boundary conditions are applied to nodes on edges *AB*, *DE* and *EFA*, respectively. A mesh is generated, using type D mesh and 8 divisions along edges *AB*, *BC* and *CD*. It consists of 281 nodes and 512 elements for an effective DOF of 1583.

Suddenly applied internal pressure is first considered. The pressure is at 300 kPa. The nonlinear elastic responses of points *A* and *B* are shown in Fig. **8.15**. These results are in very good agreement with those presented in Ref. [8.5] by Jiang and

Olson. The latter reported a peak response of 6.5 mm for both points. The vibration period was 0.004 second. On both counts, the present formulation is as accurate as the super-element employed in [8.5]. It should be mentioned that the time step size was $\Delta t = 4.5 \times 10^{-5}$ s in [8.5] while the present computation uses $\Delta t = 5 \times 10^{-5}$ s. This time step size will be used for all computations involved in this section.

Nonlinear dynamic elasto-plastic responses of points A and B at pressure levels of 100, 200 and 300 kPa are shown Fig. **8.16**. They are indicative of the tube experiencing stiffness hardening.

Suddenly applied external pressure is next considered. For nonlinear elastic responses, the pressure is kept at 300 kPa. Fig. **8.17** gives the time-histories of points A and B, and compares with [8.5]. The predicted amplitudes of vibration are close to each other, approximately 6 mm by [8.5] and 5.75 mm by the present formulation. But results of [8.5] showed a longer period of vibration. In particular, they showed that displacements bounced back to the positive (displacing inward), which the present results do not exhibit.

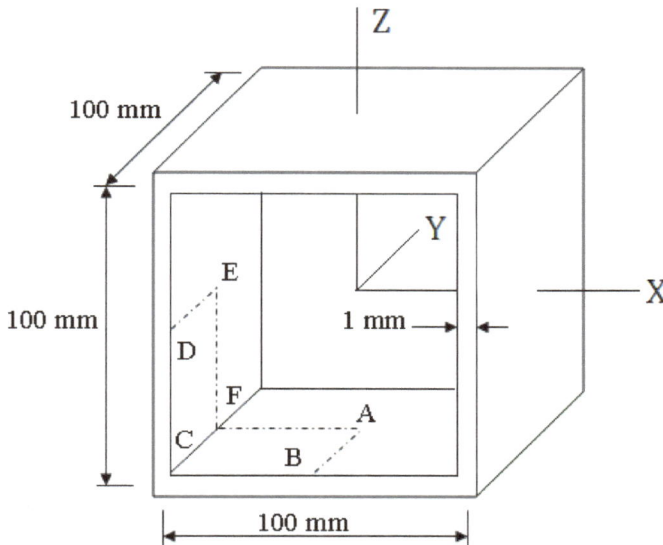

Figure 8.14: Geometry of cubic tube.

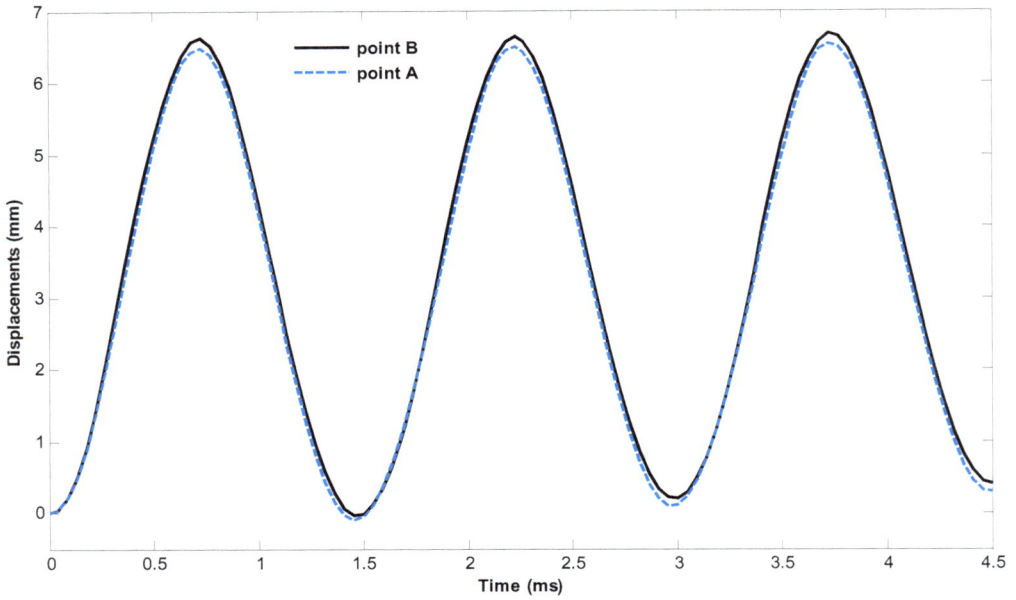

Figure 8.15: Nonlinear elastic responses of tube under internal pressure of 300 kPa.

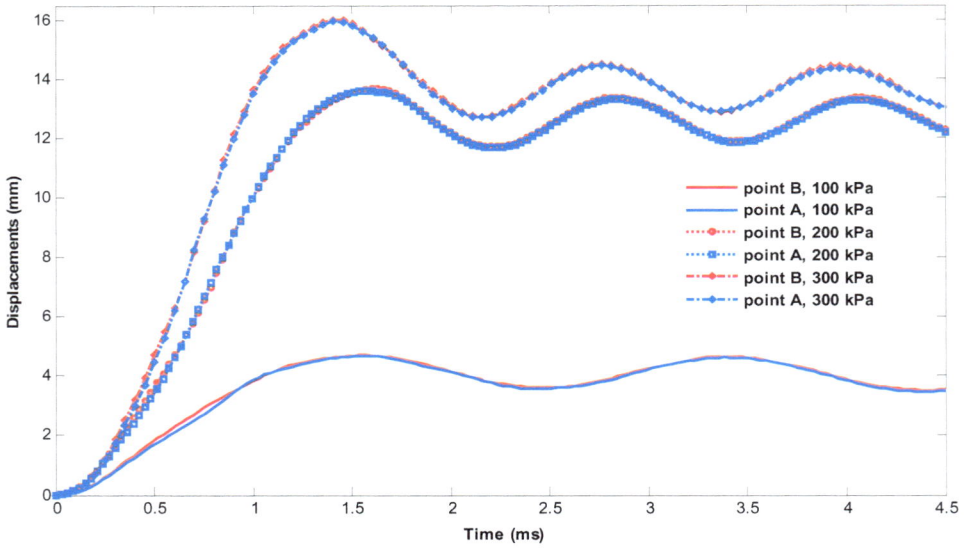

Figure 8.16: Nonlinear elasto-plastic responses of tube under internal pressure of 100, 200 and 300 kPa.

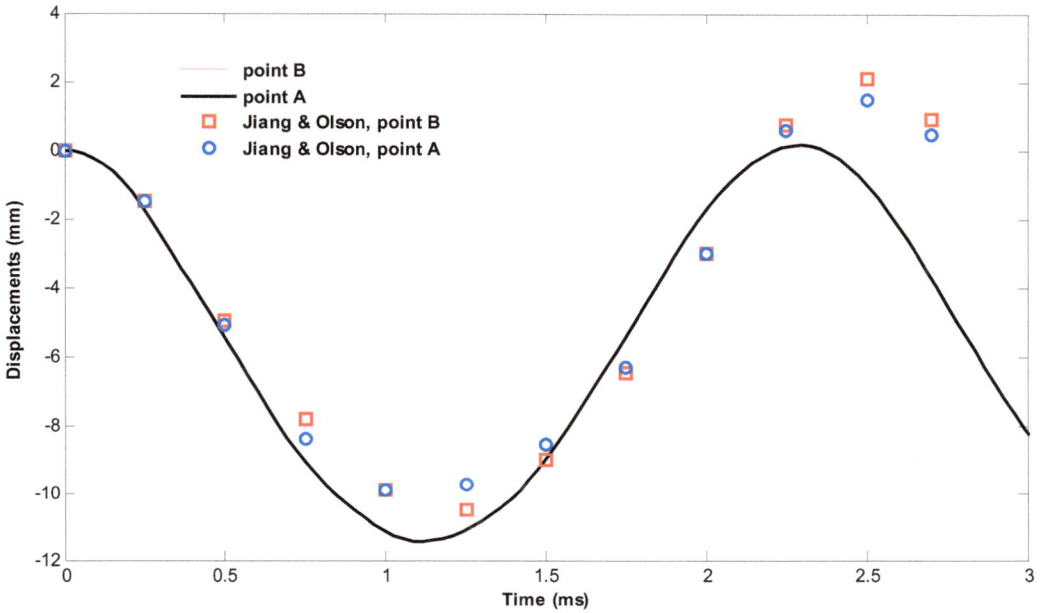

Figure 8.17: Nonlinear elastic responses of tube under external pressure of 300 kPa.

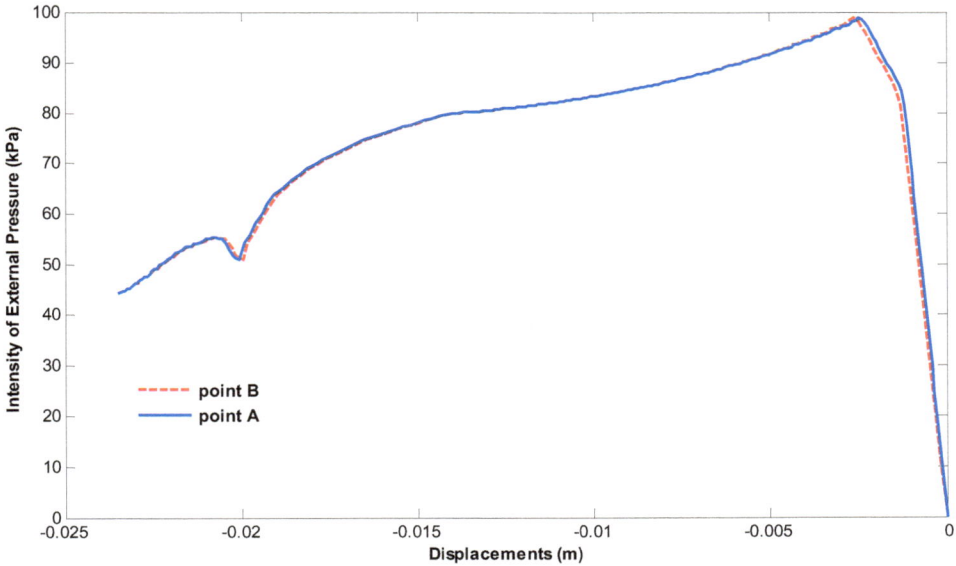

Figure 8.18: Intensity of external pressure *versus* displacements of tube.

Determining the nonlinear elasto-plastic responses due to external pressure proves more involving than the internal pressure counterpart. As shown in Fig. **8.18**, which consists of the plots of intensity of pressure *versus* displacements at *A* and *B*, the tube will sustain external pressure up to about 100 kPa. It then experiences softening. Consequently, the external pressure is set to two levels, 50 kPa and 80 kPa. Computations are carried out and time-histories of point *B* are given in Fig. **8.19**. Under the 50 kPa external pressure, the response is mainly elastic. Under the higher external pressure, the response becomes elasto-plastic. Compared with the response under 50 kPa, the period of vibration is longer and amplitude is smaller.

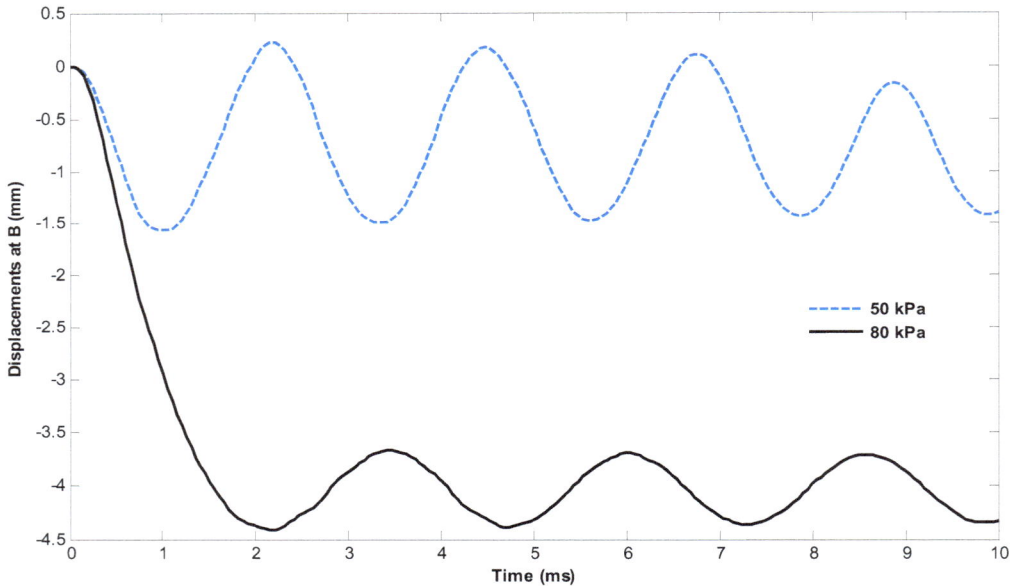

Figure 8.19: Nonlinear elasto-plastic responses of point *B* under external pressure of 50 and 80 kPa.

REFERENCES

[8.1] F.S. Almeida, and A.M. Awruch, "Corotational nonlinear dynamic analysis of laminated composite shells", *Finite Element in Analysis and Design*, vol. 47, pp. 1131-1145, October 2011.

[8.2] H.C. Huang, "Elastic and elasto-plastic analysis of shell structures using the assumed strain elements", *Comp. Struct.,* vol. 33, pp. 327-335, January 1989.

[8.3] M.A. Daye, and T.G. Toridis, "Elasto-plastic algorithms for plates and shells under static and dynamic loads", *Comp. Struct.,* vol. 39, pp. 195-205, January 1991.

[8.4] M.L. Liu, and C.W.S. To, "Vibration of structures by hybrid strain based three-node flat triangular shell elements", *J. Sound Vibr.*, vol. 184, pp. 801-821, August 1995.

[8.5] J. Jiang, and M.D. Olson, "Applications of a super element model for nonlinear analysis of stiffened box structures", *Int. J. Numer. Meth. Engrg.*, vol. 36, pp. 2219-2243, July 1993.

Send Orders for Reprints to reprints@benthamscience.net
Vibration and Nonlinear Dynamics of Plates and Shells, 2014, 167-194 **167**

CHAPTER 9

Nonlinear Dynamics of Curved-Surface Structures

Abstract: This chapter is concerned with the nonlinear dynamic responses of structures of single curvature and of double curvatures, examples of which include cylindrical shells or panels, spherical caps, and hemispheres. Geometrical nonlinearity due to large deformation, material nonlinearity due to elastic-plastic material behaviour, and various loading situation including non-conservative pressure loads, will be examined.

Keywords: Dynamics, nonlinearities, geometrical, material, spherical caps, hemispheres.

9.1. CYLINDERICAL PANEL SUBJECTED TO CENTRAL LOAD OR UNIFORM PRESSURE

In this section the nonlinear elastic response of a cylindrical panel is investigated. The panel is subjected to a central point load, or a non-conservative pressure that is uniformly distributed. The geometry of the panel is given in Fig. **9.1** where radius $R = 5$ m, length $L = 5$ m, thickness $h = 0.1$ m, and angle $\theta = 30°$. The two straight edges are simply supported and the two curved edges free. Material properties are, modulus of elasticity $E = 200$ GPa, Poisson's ratio $v = 0.25$ and mass per unit volume $\rho_0 = 10,000$ kg/m^3. A quarter of the panel is discretized by mesh10×10D. After applying appropriate boundary conditions, the resulting effective DOF is 1213.

The case of center load is first examined. The center load, $P(t)$, is a step load, see Fig. **8.10(a)**, where $t_0 = 0.2$ s and $P_0 = 50$ MN. This load level causes the panel to experience dynamic buckling [9.1, 9.2]. The nonlinear elastic response at the point of load application is shown in Fig. **9.2**. For clarity, results by Argyris *et al.* [9.1] are only included for the first 0.16 second, or the pre-buckling to buckling period. For this time period, the present formulation seems to predict a slightly more flexible behaviour of the panel than that of [9.1]. For post-buckling behaviours, the present formation yields a mean displacement of about -1.8 m with very little oscillation. On the other hand, Ref. [9.1] showed a mean displacement of approximately -1.5 m and an amplitude of approximately 0.08 to 0.09 m.

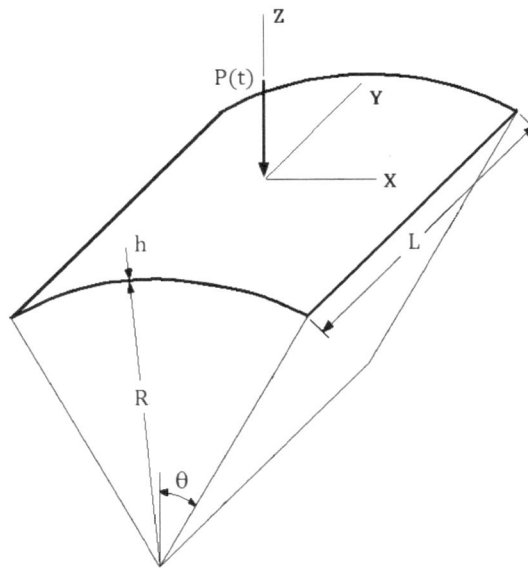

Figure 9.1: Cylindrical panel subjected to central load with straight edges being simply supported and curved edges free.

Responses due to a step, uniform, and non-conservative pressure are next investigated. The time history of the step pressure load is similar to that in Fig. **8.10(a)** in which $P(t)$ is replaced with $q(t)$. The time taken for pressure to reach full value, t_0, remains at 0.2s. Fig. **9.3** shows the nonlinear elastic responses of the center point when $q_0 = 10$ and 10.5 MPa, respectively. Note that the relationship between pressure and displacement of Fig. **9.4** suggests a nonlinear (static) buckling pressure of 10.44 MPa. If the pressure is slowly increased to, say, 10.5 MPa, the cylindrical panel would snap through to a displacement of approximately -1.4 m when the pressure reaches 10.44 MPa, and would more or less remain at that displacement when the pressure is increased further to 10.5 MPa. It may be pointed out that in this case the response below the horizontal arrow (which is applied to show the path of snap through) in Fig. **9.4** is unstable. Returning to dynamic behaviour of the panel, Fig. **9.3** demonstrates that under the 10.5 MPa pressure and post-buckling, there is very little oscillation. In contrast, the response due to the 10.0 MPa pressure exhibits clearly defined period and amplitude of vibration. It is interesting to find that -1.8 m seems to be the mean post-buckling displacements regardless of loading.

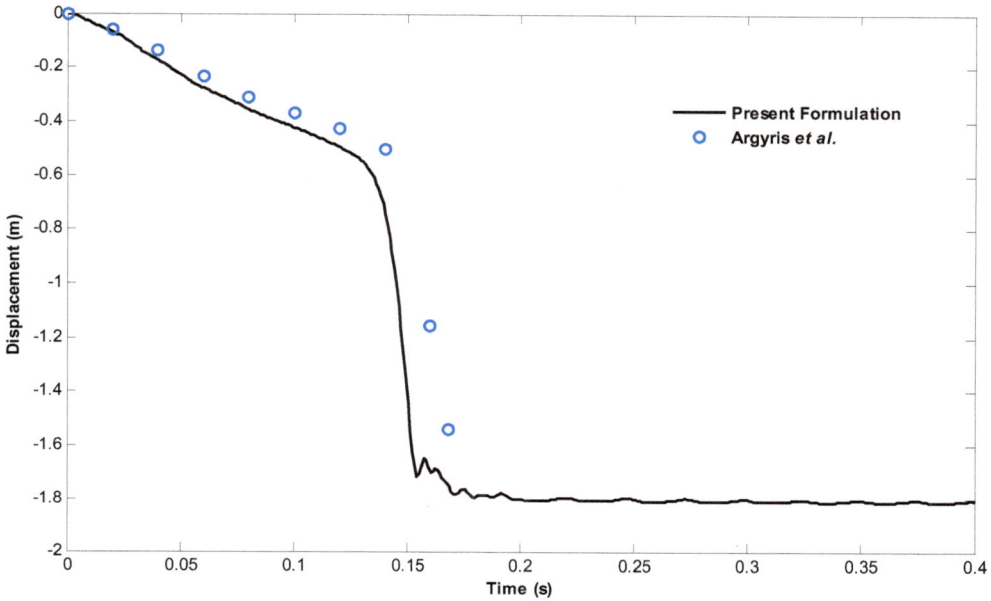

Figure 9.2: Nonlinear elastic response of cylindrical panel subjected to step center load.

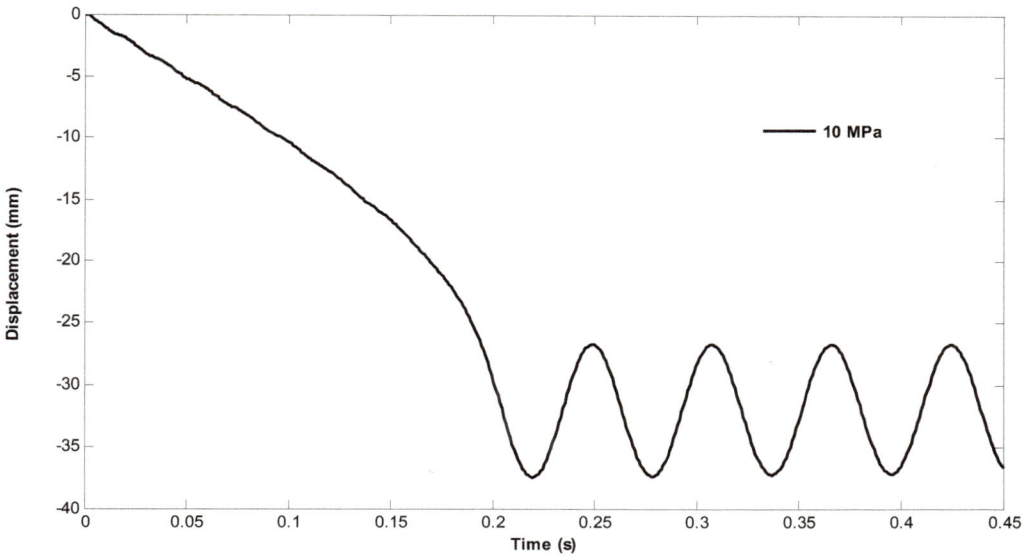

Figure 9.3(a): Nonlinear elastic response of cylindrical panel subjected to step non-conservative uniform pressure of 10 MPa.

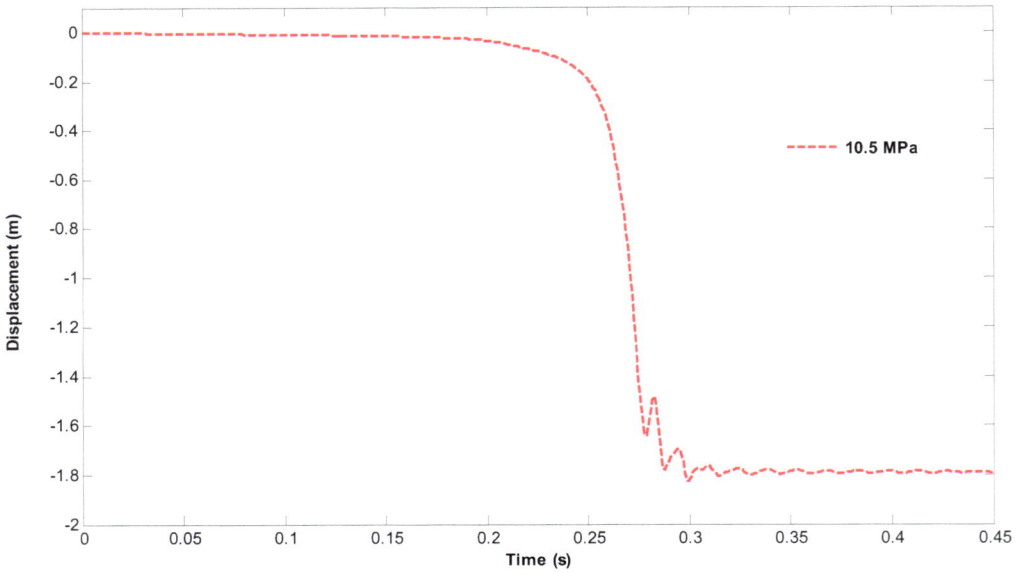

Figure 9.3(b): Nonlinear elastic response of cylindrical panel subjected to step non-conservative uniform pressure of 10.5 MPa.

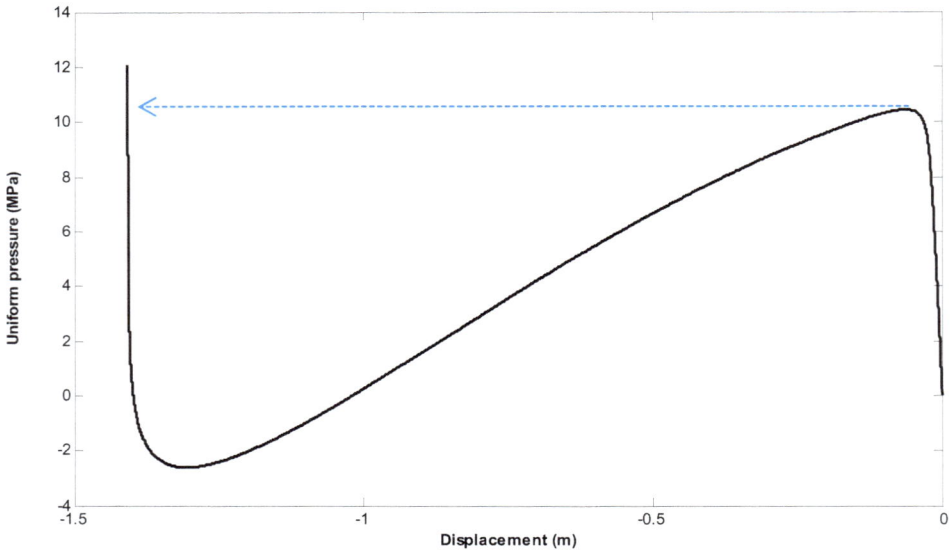

Figure 9.4: Relationship between non-conservative pressure and central displacement of cylindrical panel.

Finally, it should be pointed out that, the time step size used in the present computations is $\Delta t = 10$ ms, which is the same as in Ref. [9.1]. However, the latter employed the implicit integration scheme, the trapezoidal rule. The present investigation has experienced divergence with the trapezoidal rule but found success with the backward Euler time integration algorithm by Liu *et al.* [9.3].

9.2. SECOND CYLINDERICAL PANEL SUBJECTED TO CENTRAL LOAD

This second cylindrical panel has been investigated for its nonlinear elastic, and elasto-plastic behaviour under static loads, see [9.4], for example. With reference to Fig. **9.1**, the geometrical properties of the present panel are now: $R = 2.45$ m, $L = 0.254$ m, $h = 6.35$ mm, $\theta = 5.7296°$. The boundary conditions are the same as in Sec. 9.1. The material properties are: $E = 3.103$ GPa, $v = 0.3$, and $\sigma_Y = 1$ MPa if elastic-plastic.

The cylindrical panel exhibits snap-through phenomenon, as an elastic material and an elastic-plastic material. The load-displacement (at point of load application) curves given in Fig. **9.5** show that the snap-through load is, 300 N when material is elastic and 100 N when it is elastic-plastic. Figure **9.5** also demonstrates that, with the exception of elastic-plastic material displaced up to about 6 mm, the present results are in very good agreement with those of Montag *et al.* [9.4]. It is noted that the displacement shown in Fig. **9.5** is the absolute value of displacement.

For dynamic analyses, the mass density is set to $\rho_0 = 1150$ kg/m^3. Suddenly applied center load is assumed similar to that shown in Fig. **8.5** with $q(t)$ being replaced by $P(t)$. The load level, P_0, is chosen such that dynamic snap-through will take place. Hence, the following initial values: $P_0 = 300$ N for elastic material and $P_0 = 100$ N for elastic-plastic material. Computed nonlinear dynamic responses at the point of load application are shown in Figs. **9.6** and **9.7**. For the case of elastic material (Fig. **9.6**), the panel behaves, at load levels of $P_0 = 300$ and 330 N, in a way very similar to the cylindrical panel investigated in the previous section. When the material is elastic-plastic (Fig. **9.7**), the present result suggests a snap-through of the panel to about -2.7 cm or -27 mm. Post-buckling response is not available due to singularity. At a reduced load of 90 N, the panel seems to be displaced to about -2.65 mm and to remain at that displacement afterward.

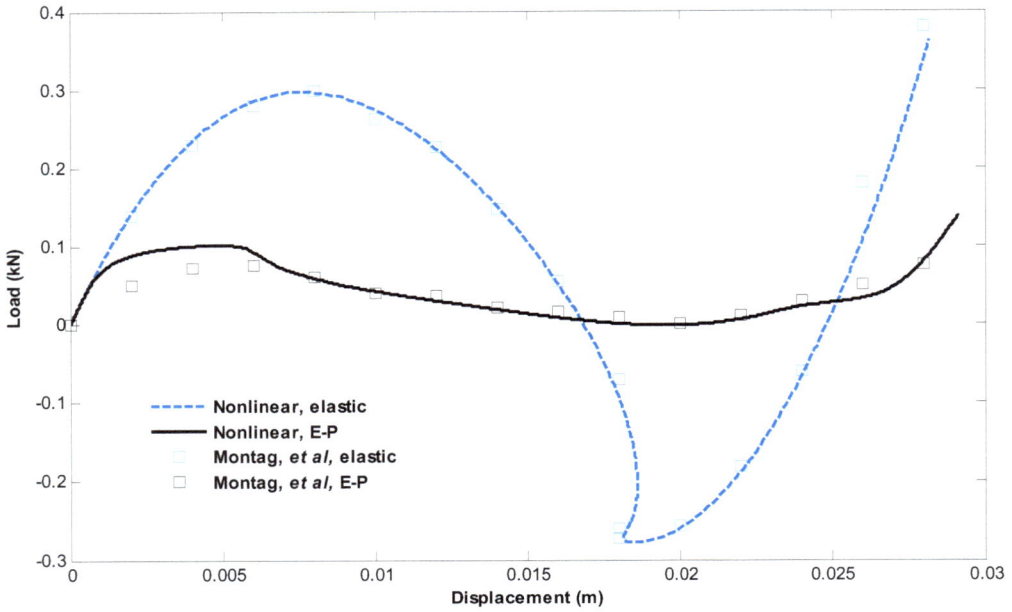

Figure 9.5: Relationship between load and displacement of cylindrical panel.

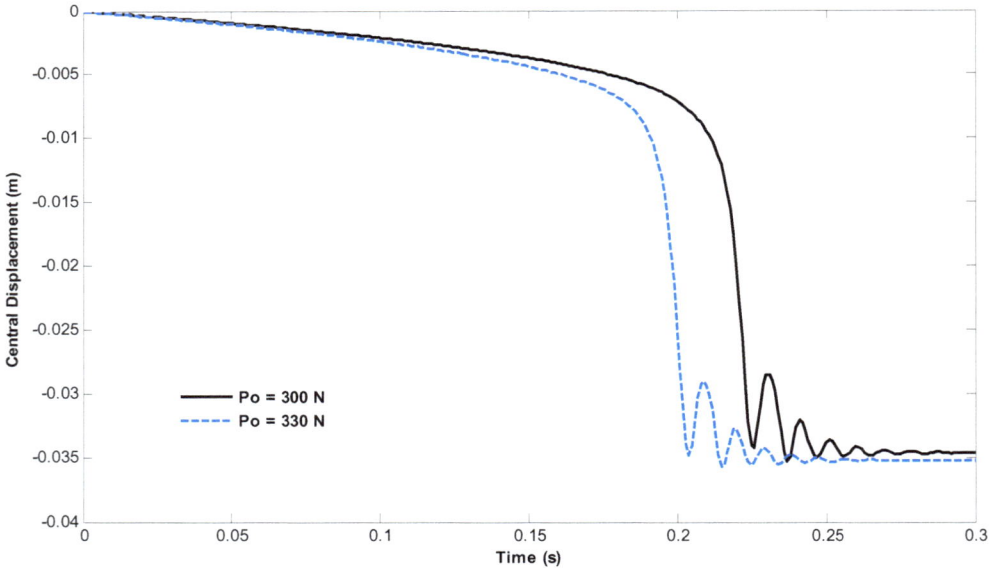

Figure 9.6: Nonlinear elastic responses of cylindrical panel.

In passing, it should be mentioned that the mesh used in the computations in this section is 12×12C which discretizes one quarter of the panel and has an effective DOF of 883. The time step size adopted for the dynamic analyses is $\Delta t = 1.0$ ms. The backward Euler scheme of [9.3] is again employed.

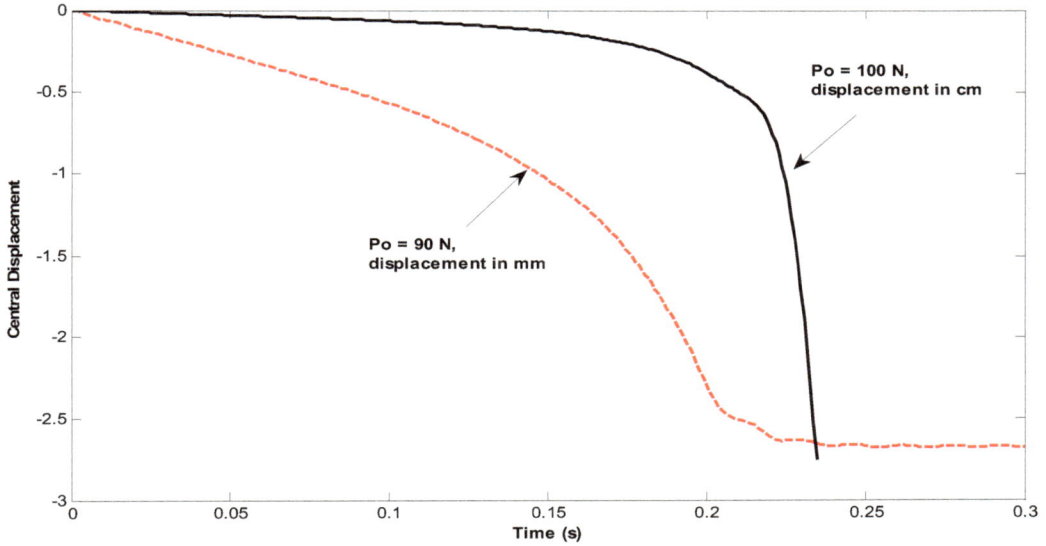

Figure 9.7: Nonlinear elasto-plastic responses of cylindrical panel.

9.3. HEMISPHERE WITH CENTRAL HOLE AND SUBJECTED TO ALTERNATING LOADS

The geometry, boundary conditions and loading on one quadrant of the hemisphere are depicted in Fig. **9.8**, where $R = 254$ mm (10 in), $h = 1.106$ mm (0.04 in) and $\theta = 18°$. The material is elastic-perfectly plastic with properties of $E = 470.6$ GPa ($68.26×10^6$ psi), $v = 0.3$, $\sigma_Y = 689.5$ MPa (100 ksi), and $\rho_0 = 10,685$ kg/m^3 (0.001 lb·s^2/in^4). These values are adopted from Key and Hoff [9.5] for easy comparison. A 10×10D mesh is employed to discretize the quadrant. It has 221 nodes, 400 elements and an effective DOF of 1260. The fundamental period of the corresponding linear elastic system is 44.91 ms. Two alternating point loads are applied at points A and B, respectively. It should be noted that $P(t)$ is the force exerted on the quadrant.

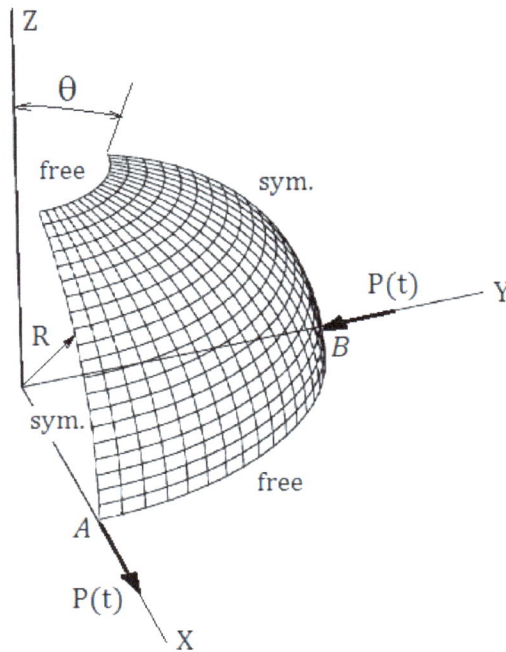

Figure 9.8: Geometry, boundary conditions and loads of hemisphere.

Reference [9.5] examined the linear dynamic response of the hemisphere, assuming suddenly applied loads which are similar in form to that in Fig. **8.5** with $q(t)$ being replaced by $P(t)$ such that $P_0 = 4.45$ N (1 lb). In Fig. **9.9**, the presently computed linear dynamic response is compared with that of [9.5]. The latter predicted a higher peak value (approximately 4.75 mm) and longer period (approximately 48.0 ms) of oscillation. The present formulation yields a smaller peak value (4.5 mm) and shorter period of vibration (45.0 ms). The discrepancies can be reduced by employing finer meshes.

The nonlinear elastic response of point A due to the same suddenly applied point loads, is also included in Fig. **9.9**. It is observed that, at the given load level, the hemisphere seems to experience some stiffness softening. It should be mentioned that, the response at point B is not shown in Fig. **9.9** because at such a load level the responses of A and B are opposite to each other. However, this is no longer the case at elevated load levels and large deformation.

Figs. **9.10** and **9.11** show the nonlinear elastic responses of points A and B at various load levels. Both locations see their periods of oscillation slightly increased then decreased as P_0 is increased. Meanwhile, the mean displacements at the points suggest that they drift outward at A or inward at B as time goes on. This drifting becomes more pronounced at higher load level. It is also interesting to observe that, with respect to increasing time, the amplitudes of vibration at A decrease while those at B increase.

For nonlinear elasto-plastic analysis, step loads whose time history is shown in Fig. **8.10(a)** are considered. The rise time or time of the ramp function is $t_0 = 0.6$ ms, and P_0 is set to 133.5 N (30 lb). The computed nonlinear elasto-plastic responses are given in Fig. **9.12**. It is seen that, under the step loads, both points oscillate at the same frequency but with different amplitudes. Further, both points experience stiffness softening in the elasto-plastic regime. It should be noted that the displacements at B are shown in their absolute values.

Finally, the time step size used in the computations is 0.8 ms. Results in Figs. **9.9** through **9.12** are obtained by employing the backward Euler time integration scheme proposed by Liu *et al.* [9.3].

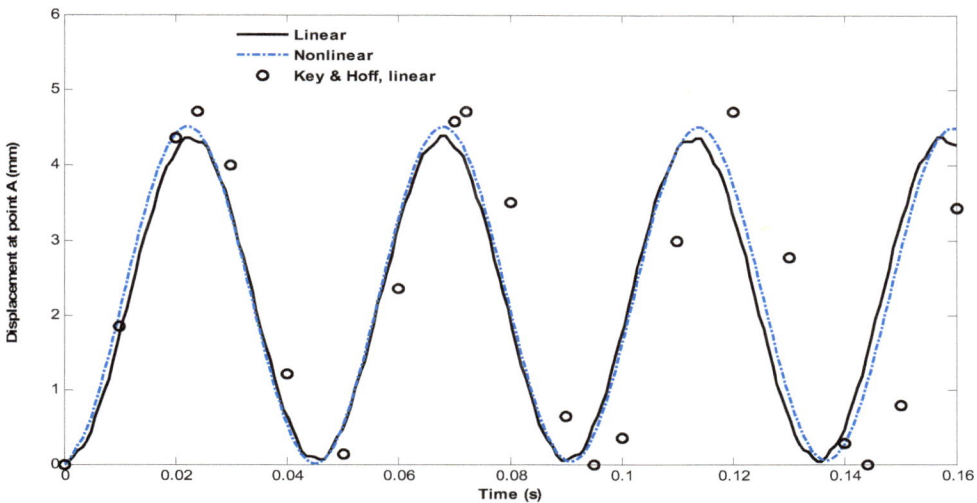

Figure 9.9: Linear and nonlinear elastic responses at point A of hemisphere to suddenly applied loads.

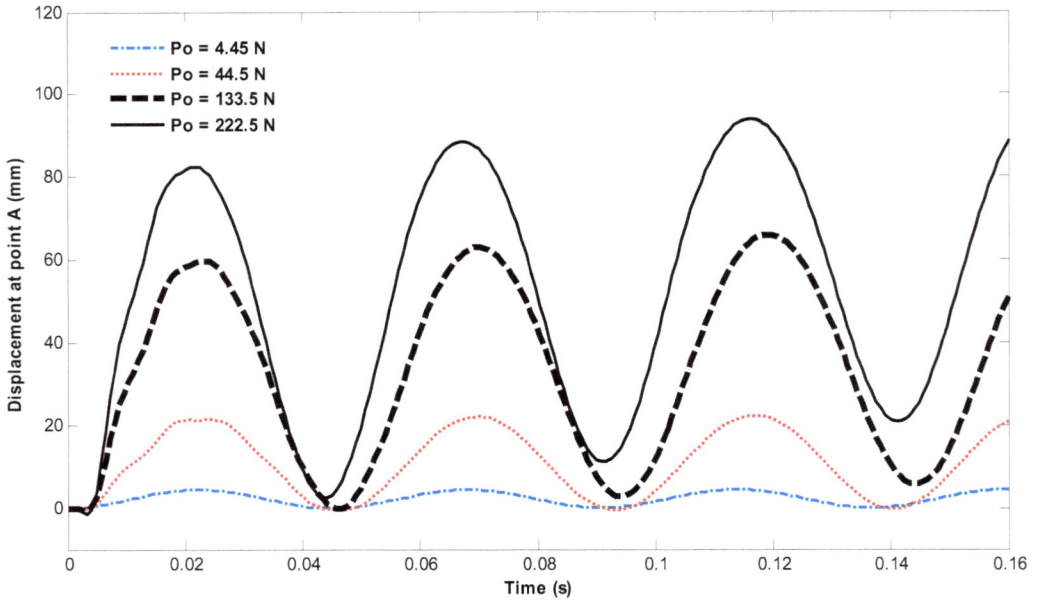

Figure 9.10: Nonlinear elastic responses at point *A* of hemisphere to suddenly applied loads.

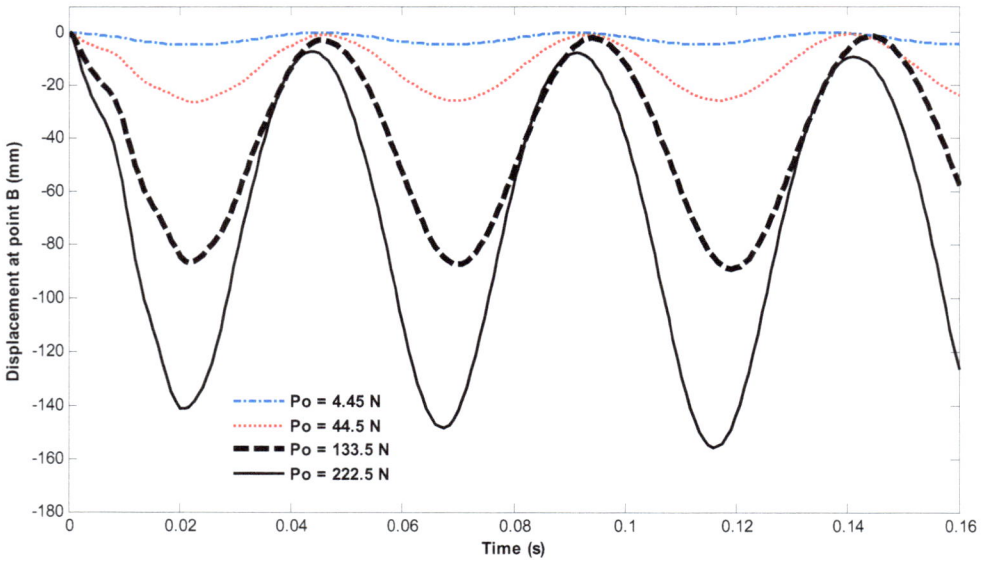

Figure 9.11: Nonlinear elastic responses at point *B* of hemisphere to suddenly applied loads.

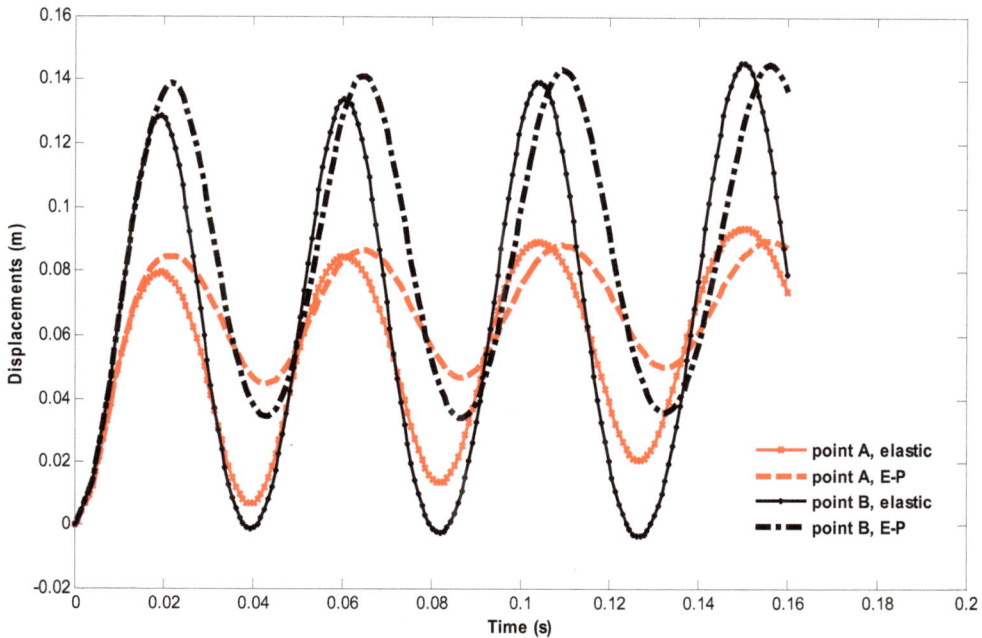

Figure 9.12: Nonlinear elastic and elasto-plastic responses of hemisphere to step loads.

9.4. HEMISPHERE SUBJECTED TO ALTERNATING POINT LOADS

Figure **9.13** pictures one quadrant of the hemisphere. Its geometrical properties are, $R = 100$ mm and $h = 5$ mm. The material is elastic-plastic with linear isotropic hardening. The material properties are: $E = 100$ MPa, $v = 0.2$, the tangent modulus of elasticity $E_T = 23.08$ MPa, $\sigma_Y = 2.0$ MPa, and $\rho_0 = 1000.0$ kg/m^3. The quadrant is represented by a 370D mesh which has 201 nodes, 370 elements and an effective DOF of 1141. The fundamental period of the corresponding linear elastic system is 31.4 ms. The two alternating point loads at points A and B, respectively, are forces exerted on the quadrant. They are the step loads that have the similar form given in Fig. **8.10(a)**. The parameters in the latter are, $t_0 = 0.02$ s, and $P_0 = 10.0$ and 20.0 N, respectively. Computed nonlinear elasto-plastic responses at points A, B and C are shown in Fig. **9.14**. All three points exhibit stiffness hardening when $P(t)$ is increased. The responses are computed using the algorithm of [9.3] with a time step size of $\Delta t = 1.5$ ms.

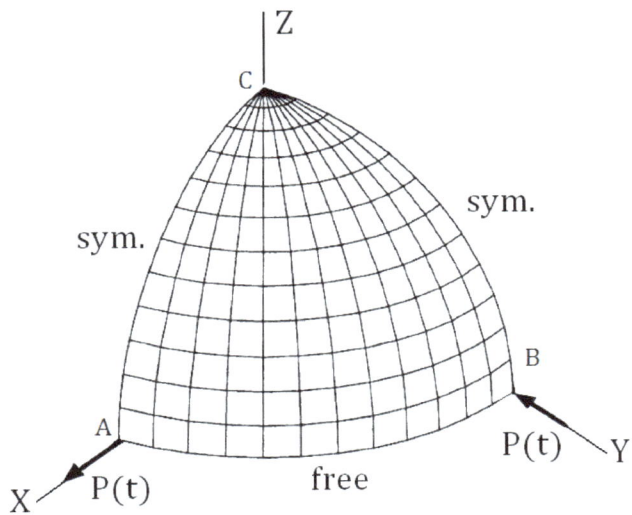

Figure 9.13: Geometry, boundary conditions, and loads on hemisphere.

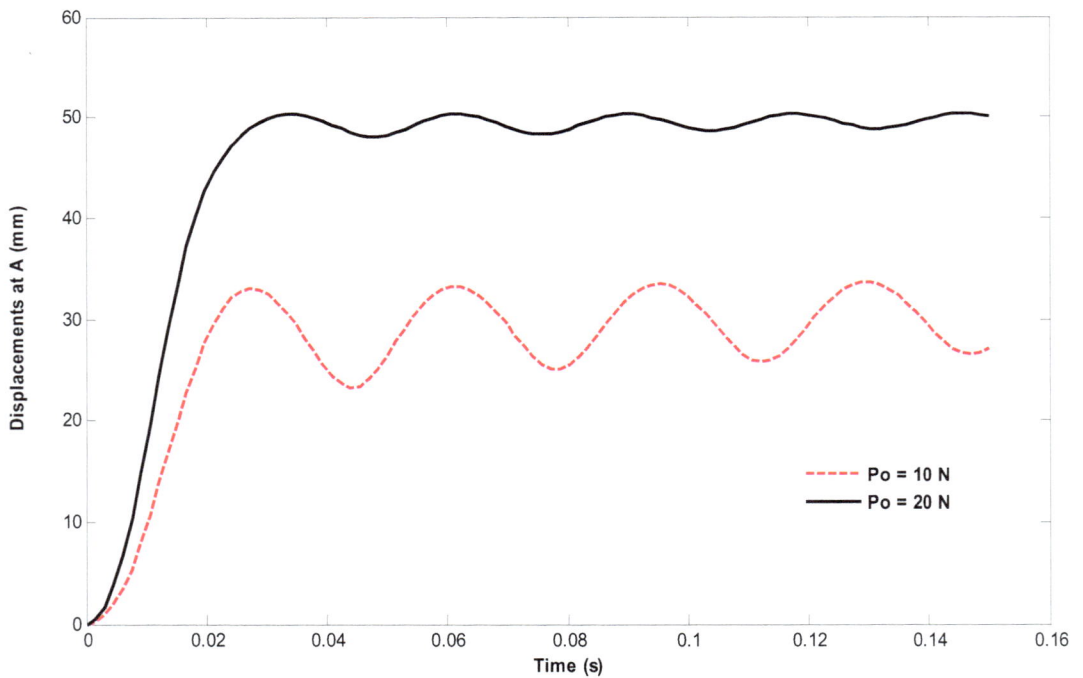

Figure 9.14(a): Nonlinear elasto-plastic responses at point *A* of hemisphere to step loads.

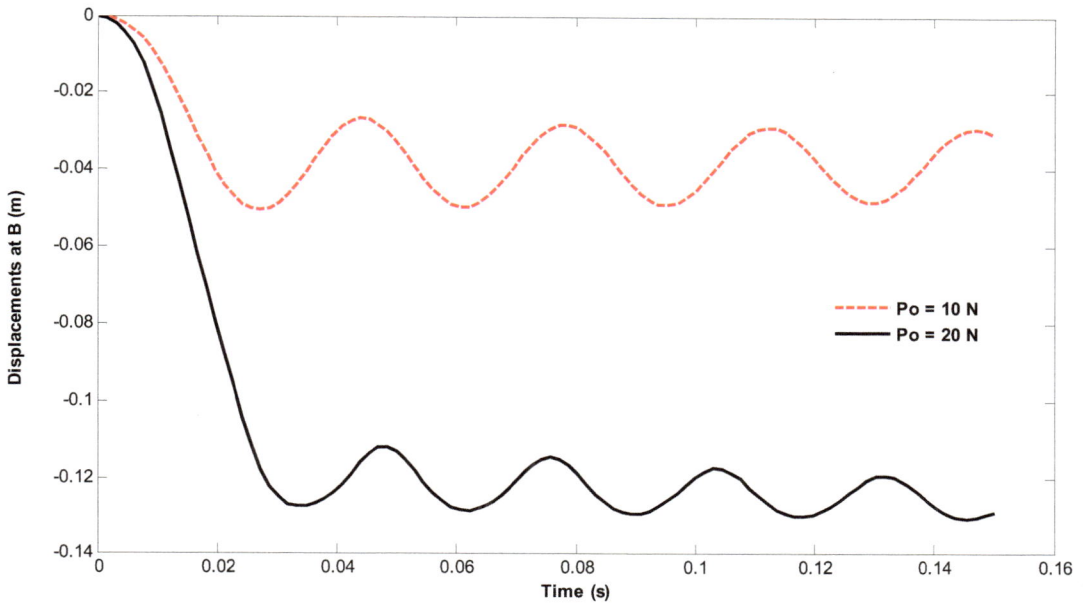

Figure 9.14(b): Nonlinear elasto-plastic responses at point B of hemisphere to step loads.

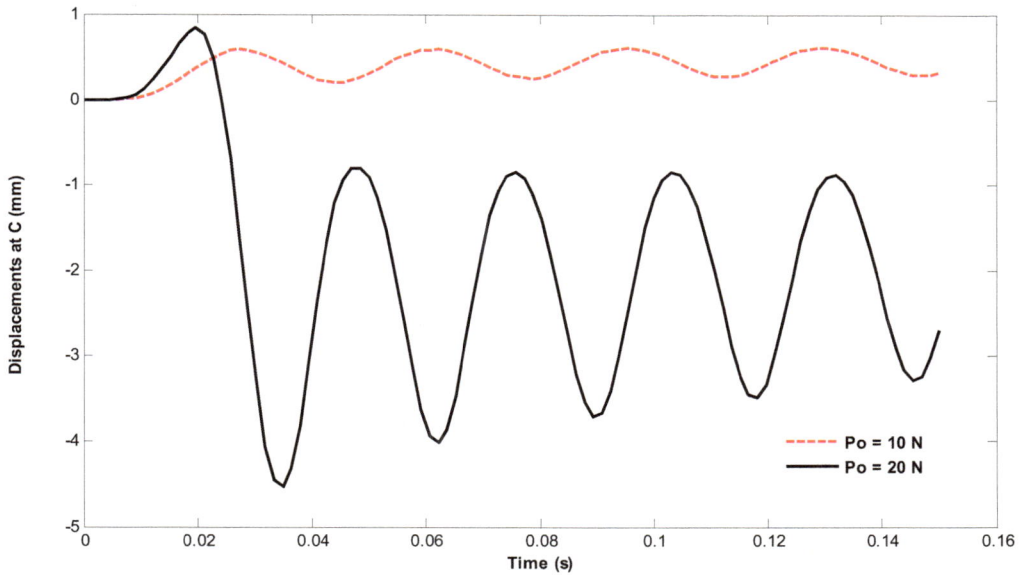

Figure 9.14(c): Nonlinear elasto-plastic responses at point C of hemisphere to step loads.

9.5. CLAMPED SPHERICAL CAP TO UNIFORM PRESSURE

The spherical cap examined in this section is shown in Fig. **9.15** where $R = 494.9$ mm, $h = 1.0$ mm and $\theta = 7.13°$. The material is elastic with $E = 200.0$ GPa, $v = 1/3$ and $\rho_0 = 7850.0$ kg/m^3. A 200D mesh is employed to represent one quarter of the cap. The mesh has 121 nodes, 200 elements and an effective DOF of 541. Polat and Calayir [9.6] found the dynamic axi-symmetric snap-through pressure to be at 0.62 MPa. For nonlinear dynamic responses, suddenly applied uniform pressure is considered and illustrated in Fig. **8.5**, with the pressure level $q_0 = 0.5$ MPa. Nonlinear elastic responses of the cap under pressure are given in Fig. **9.16**. They are compared with results by Polat and Calayir [9.6]. In the present computations, both conservative and non-conservative pressures are considered since such information was not provided explicitly in [9.6]. It is seen that the response due to non-conservative pressure lags behind that due to conservative pressure, but exhibits very similar overall behaviours. Compared with results of [9.6], discrepancy is seen. It may be attributed to a number of factors. For example, Ref. [9.6] employed an iso-parametric axi-symmetric total Lagrangian (TL) based shell element, and the trapezoidal rule of direct integration with a time step size of $\Delta t = 7.5$ μs. On the other hand, the present computations are carried out by the backward Euler time-integration algorithm by Liu *et al.* [9.3] with $\Delta t = 10.0$ μs. The present results show smaller periods of vibration.

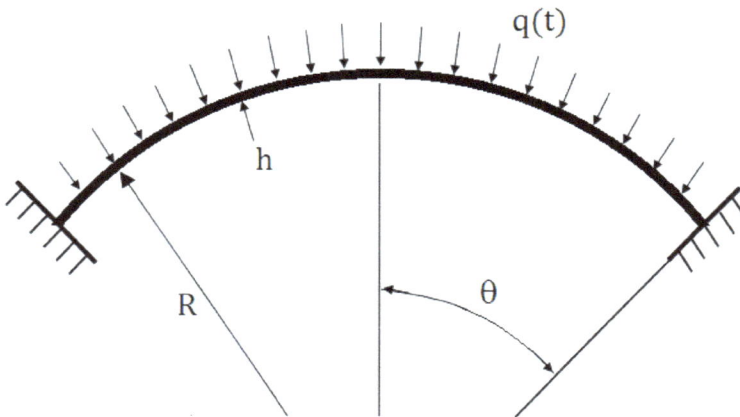

Figure 9.15: Clamped spherical cap to uniform pressure.

To examine the ability of the present formulation to deal with snap-through buckling, the pressure levels are varied between 0.8 MPa and 1.2 MPa. They are chosen, based on the knowledge that, the linear (static) buckling pressure is $q_{cr} = 2Eh^2 / \left[R^2 \sqrt{3\left(1 - \nu^2\right)} \right] = 1.0$ MPa and the nonlinear (static) buckling pressure is 0.6315 MPa, with reference to Fig. **9.17** and nonlinear dynamic buckling pressure is 0.62 MPa [9.6]. The computed nonlinear elastic responses of the cap under non-conservative pressures of 0.5, 0.8 and 1.2 MPa are presented in Fig. **9.18**. It is observed that as the uniform pressure is increased, the period and amplitude of oscillation of the apex displacement get smaller, an indication of stiffness hardening. Under pressure loads of 0.8 and 1.2 MPa, the apex displacement experiences a rapid increase in its magnitude in the first 0.5 ms or less, indicative of dynamic snap-through. On the other hand, the response under the 0.5 MPa uniform pressure shows a slower process toward dynamic snap through, taking about 1.5 ms. Note that at a displacement of -6 mm which is the mean displacement at the apex post-buckling, the stiffness is relatively small (see Fig. **9.17**), hence the larger period and amplitude of oscillation under a pressure of 0.5 MPa.

In passing, it should be mentioned that, unlike the cylindrical panels investigated in Sec. 9.1 and 9.2 which show very little post-buckling oscillation, the clamped spherical cap exhibits substantial post-buckling oscillation.

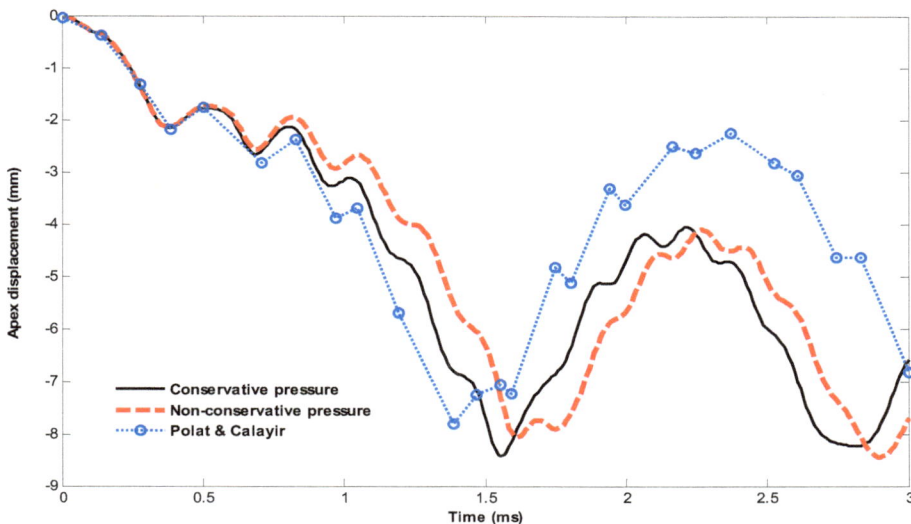

Figure 9.16: Nonlinear elastic responses of clamped spherical cap under pressure of 0.5 MPa.

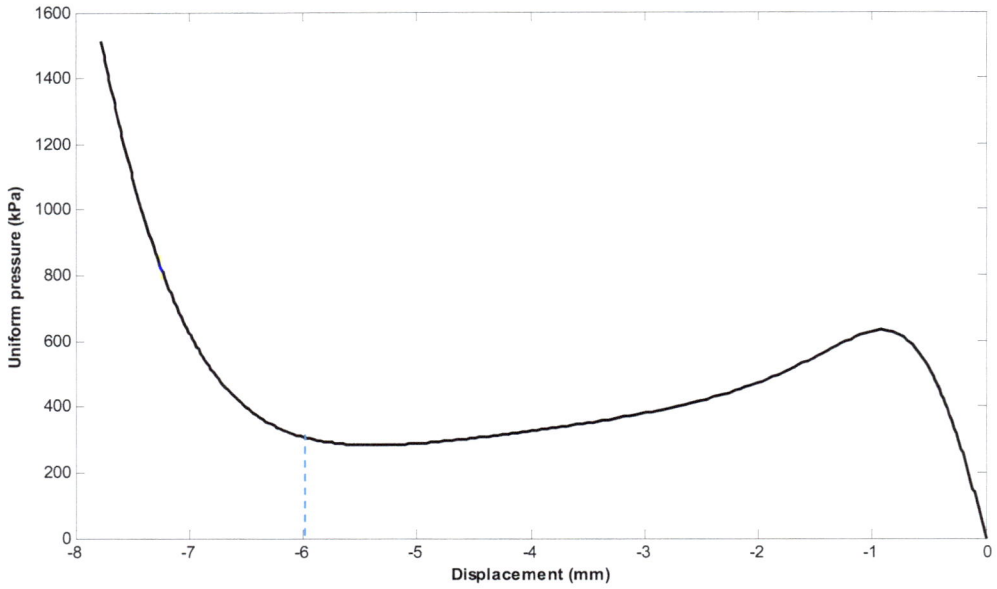

Figure 9.17: Relationship between of non-conservative pressure and apex displacement.

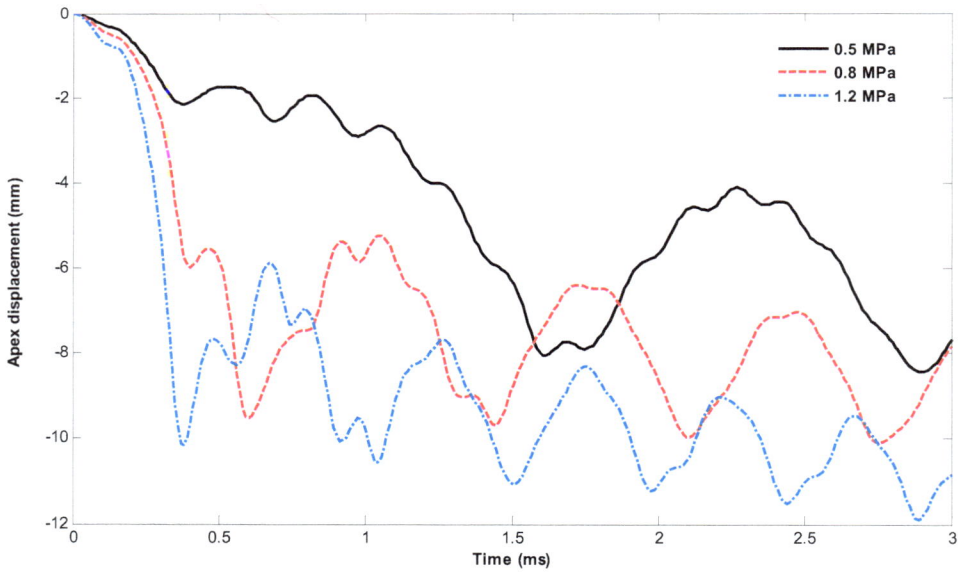

Figure 9.18: Nonlinear elastic responses of clamped spherical cap under non-conservative pressures.

9.6. CLAMPED SPHERICAL CAP TO APEX LOAD

The spherical cap that is investigated in this section is shown in Fig. **9.19**. This clamped spherical cap has the following geometrical and material properties: radius R = 120.90 mm (4.76 in), thickness h = 0.4 mm (0.01576 in), sagitta H = 2.18 mm (0.08589 in), θ = 10.9°, Young's modulus E = 68.95 GPa (10^7 psi), Poisson's ratio v = 0.3, yield strength σ_Y = 344.75 MPa (50 ksi) which was stated incorrectly in [9.7] as 139.7 MPa (20 ksi), tangent modulus E_T = 0, and mass density ρ_0 = 2651.5 kg/m^3 (0.000245 lb·s^2/in^4). The apex loads $P(t)$ are of the step type similar to that shown in Fig. **8.5** in which $q(t)$ is replaced by $P(t)$ such that P_0 = 155.89 N (35 lb), and an impulse type similar in form to that given in Fig. **8.10(b)** except that the rise time t_0 = 1.0 ms, the duration for the constant load is 0.5 ms, and the load reduction time is also 0.5 ms. The peak load P_0 = 133.5 N (30 lb). Two different meshes for one quadrant of the cap are considered in the following two sub-sections. The first mesh model has been studied by the authors in [9.7] and is included in Sub-section 9.6.1 whereas the results of the second mesh model are examined in Sub-sec. 9.6.2.

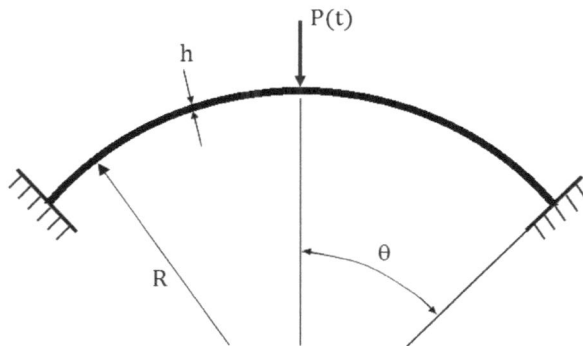

Figure 9.19: Clamped spherical cap subjected to apex load.

9.6.1. Coarse Mesh Model

This model has 46 nodes and 72 elements. It is reproduced from [9.7] and included in Fig. **9.20**. Note that more finite element models of coarser and finer meshes can be found in the latter reference but not presented in this sub-section for brevity. It suffices to mention that for static loads all these meshes gave accurate results compared with those reported by Oliver and Onate [9.8].

Computed results applying the shell elements described in Chapter 7 for the case with step load are reproduced from [9.7] and presented in Fig. **9.21** while computed results for the case with impulse load are reproduced from [9.7] and included in Fig. **9.22**. In both Figs. **9.21** and **9.22** computed results applying the computer program by Bathe *et al.* [9.9] were included for comparison. In these results the trapezoidal rule for numerical integration was applied with a time step size $\Delta t = 2.2$ μs. The latter is approximately 1/50 of the fundamental period of the clamped spherical cap which is 0.116 ms [9.9].

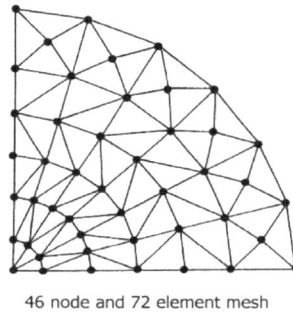

46 node and 72 element mesh

Figure 9.20: Coarser finite element model of quadrant spherical shell [9.7].

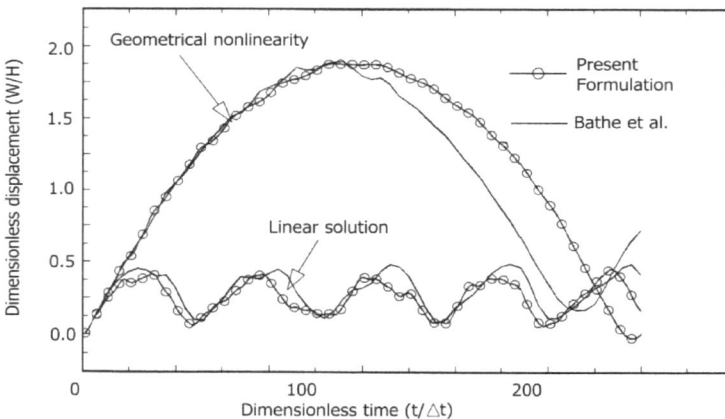

Figure 9.21: Apex dynamic response of clamped spherical shell under step load [9.7].

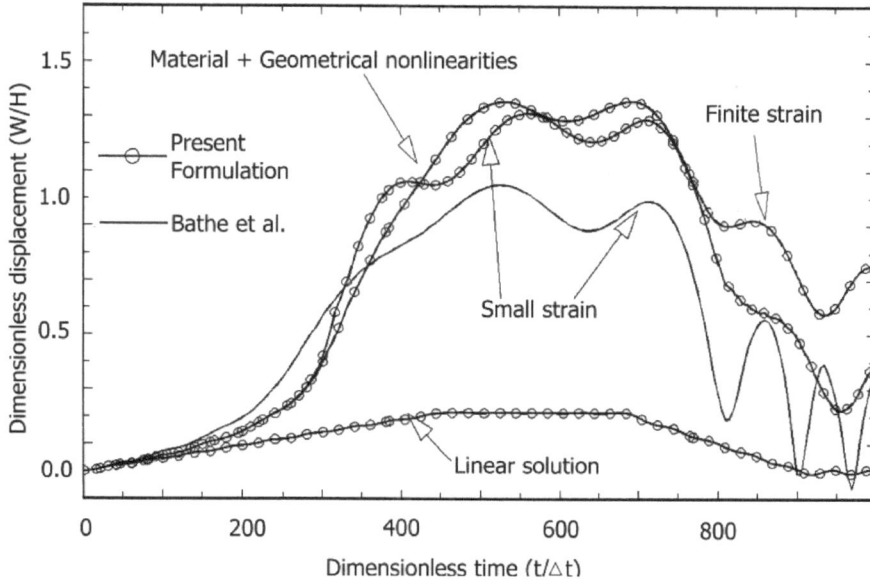

Figure 9.22: Apex dynamic response of clamped spherical shell under impulse load [9.7].

9.6.2. Refined Mesh Model

The quadrant spherical cap represented by the 200D mesh in Sec. 9.5 is applied in this sub-section. The geometrical and material properties are identical to those for the coarser mesh model. The impulse load whose time history is similar in form to that in Fig. **8.10(b)** except that the rise time now is $t_0 = 1.0$ ms, the duration for the constant load portion is 0.5 ms, and the time corresponding to the load reduction part is 0.5 ms. The peak $P_0 = 133.5$ N (30 lb) is applied at the apex. Computed nonlinear elasto-plastic responses at the apex are shown in Fig. **9.23** where they are compared with the responses computed by applying the computer program by Bathe *et al.* [9.9]. Very good agreement is seen for the first 1.8 ms. It is also observed that, compared with the coarse mesh of Fig. **9.20**, the refined 200D mesh yields responses closer to those by [9.9], which is expected, due to the upper bound nature of the formulation.

Subsequently, the yield strength of the material is set to a lower value of $\sigma_Y = 137.9$ MPa (20 ksi) which was incorrectly reported as 139.7 MPa in [9.7]. Four different load levels are chosen. They range from lower to higher than the nonlinear buckling load. Figure **9.24** gives the relationship between load and displacement of the cap in which the material undergoes elastic-plastic deformation. Computed nonlinear elasto-plastic responses under the chosen load levels are presented in Fig. **9.25**. With reference to the computational experiments performed during the course of the present investigation, it is seen that the nonlinear buckling load is about 62.75 N (14.1 lb). On the other hand, the nonlinear dynamic responses show that, under the impulse apex load which is lower than that causing snap-through, the apex experiences some peak displacement but "settles unto" small amount of permanent displacement. When the applied center load is increased beyond the buckling load, the apex seems to take time to snap-through. At the load level of P_0 = 133.5 N (30 lb) post-buckling oscillation is visible, albeit to a lesser extent. For this load level after the first 2.0 ms which is the duration of the applied impulse, the displacement response is time-independent. The computations in this sub-section have been performed using the backward Euler algorithm by Liu *et al.* [9.3] with Δt = 11.0 μs which is 5 times of that using the trapezoidal rule in Sub-sec. 9.6.1. Note that displacements shown in Figs. **9.23** through **9.25** are their absolute values.

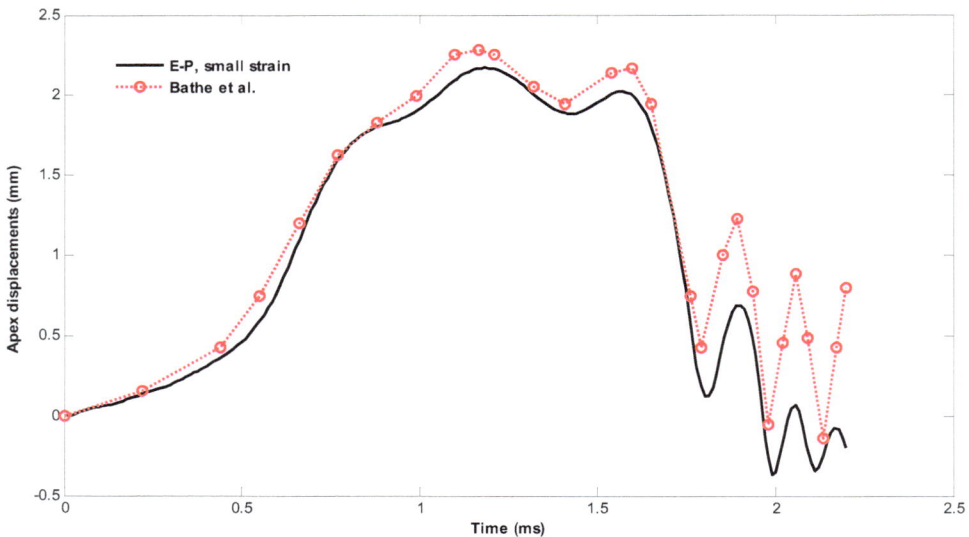

Figure 9.23: Nonlinear elasto-plastic responses of clamped spherical cap.

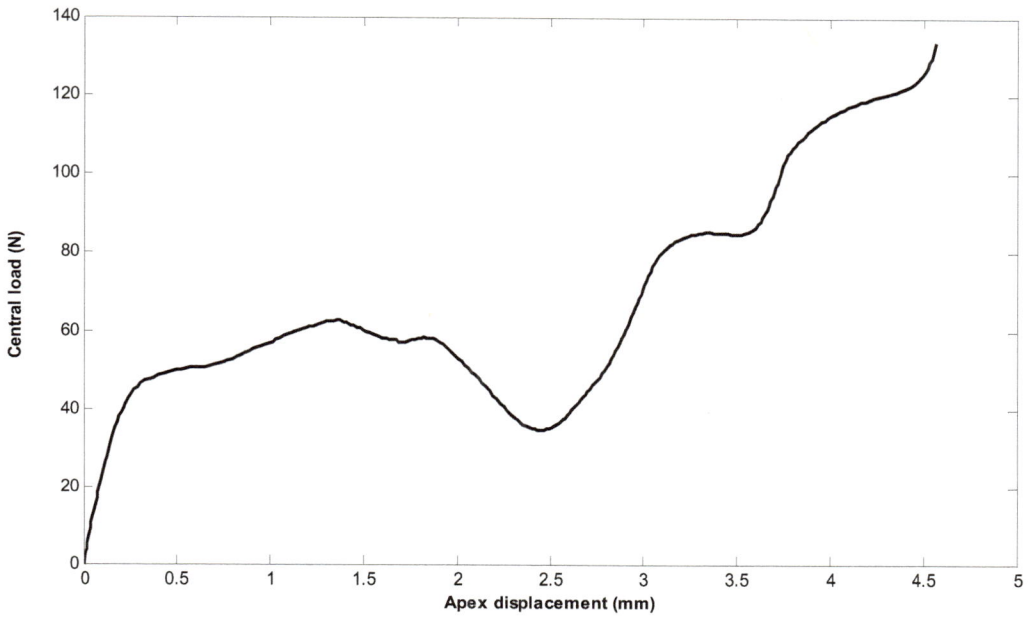

Figure 9.24: Relationship between load and elastic-plastic displacement of clamped spherical cap.

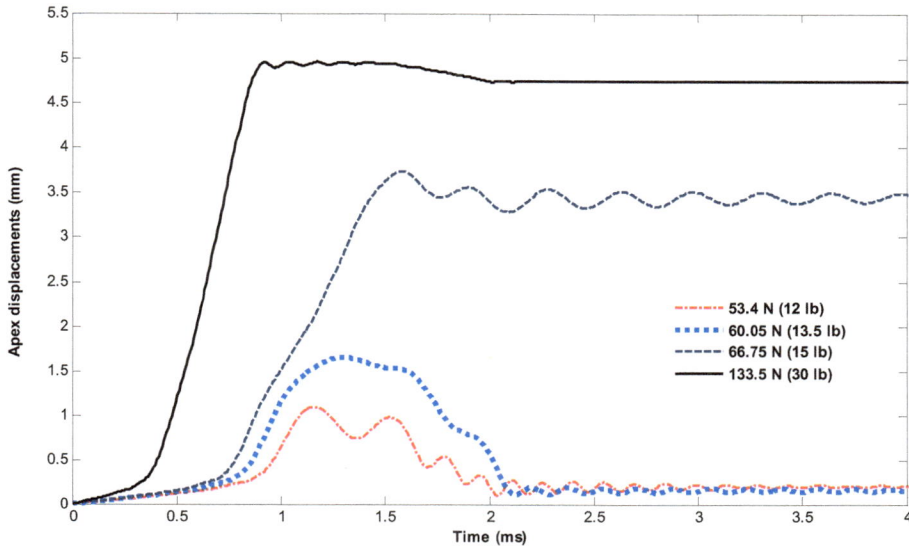

Figure 9.25: Nonlinear elasto-plastic responses of clamped spherical cap.

9.7. HINGED SPHERICAL CAPS

In this last section of the chapter, the clamped spherical caps that have been investigated in Sec. 9.5 and 9.6 are further examined for the hinged or simply supported boundary condition. These spherical caps have the identical geometrical and material properties as those provided in Sec. 9.5 and 9.6, unless stated otherwise.

9.7.1. Uniform Pressure

The simply-supported spherical cap is under a uniform pressure. The present 200D mesh that is employed to represent one quarter of the cap has 121 nodes and 200 elements. Thus, it has an effective DOF of 600. For nonlinear dynamic responses, suddenly applied uniform pressure is considered and illustrated in Fig. **8.5**, with the pressure level $q_0 = 0.375$ and 0.75 MPa, respectively. Nonlinear elastic responses at the apex are given in Fig. **9.26**. Both conservative and non-conservative pressures are considered. Relatively small difference in dynamic response is observed between conservative and non-conservative pressures, and at both pressure levels. Fig. **9.27(a)** shows the relationship between load and nonlinear elastic displacement of the spherical cap, which demonstrates nearly identical behaviours up to a pressure level of 1.0 MPa. Considering the case in which the pressure, conservative or non-conservative, is increased from zero, or from point A in Fig. **9.27(a)**, the loading path AB is stable. From B unloading follows the path of BCD which is unstable. Loading would then take place again, following the stable path of DE. The path beyond E is again unstable. The deformed configurations of the cap at B, C, D and E are shown in Figs. **9.27(b)** through **9.27(e)**. They demonstrate that, pre-buckling, the cap deforms in the axisymmetric way. Once the buckling pressure (or limit point B) is reached, non-axisymmetric deformation develops, which persists through to E and perhaps beyond. For clamped spherical caps subjected to uniform pressure, Refs. [9.10, 9.11] reported similar post-buckling path.

It should be mentioned that the configurations shown in Figs. **9.27(b)** through **9.27(e)** are not plotted in scale. To indicate the variation in displacement, color map is added based on displacement W. Consequently, the dark red color indicates the maximum displacement while the deep blue indicates the minimum displacement.

Finally, it is noted that the present computations have been performed by applying the backward Euler time-integration algorithm by Liu *et al.* [9.3] with $\Delta t = 10$ μs.

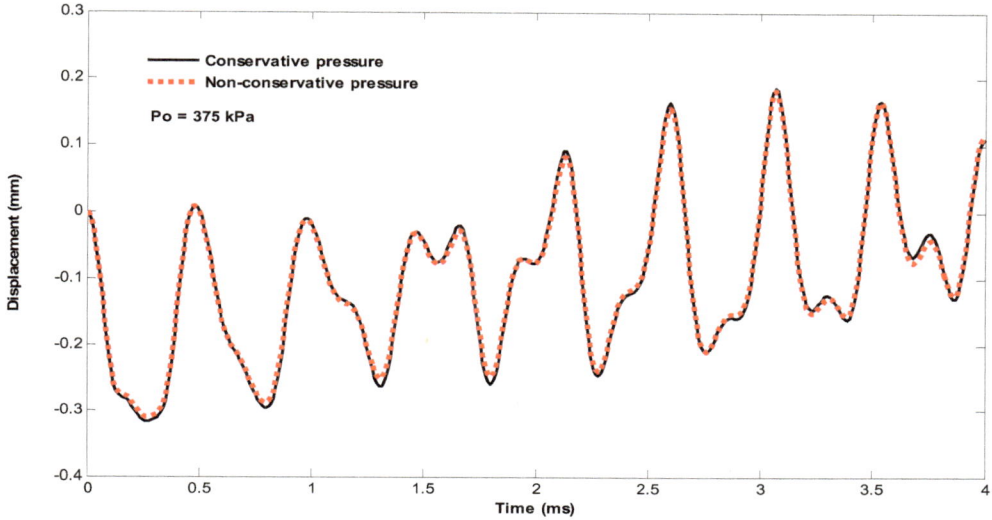

Figure 9.26(a): Nonlinear elastic responses of spherical cap at $P_0 = 375$ MPa.

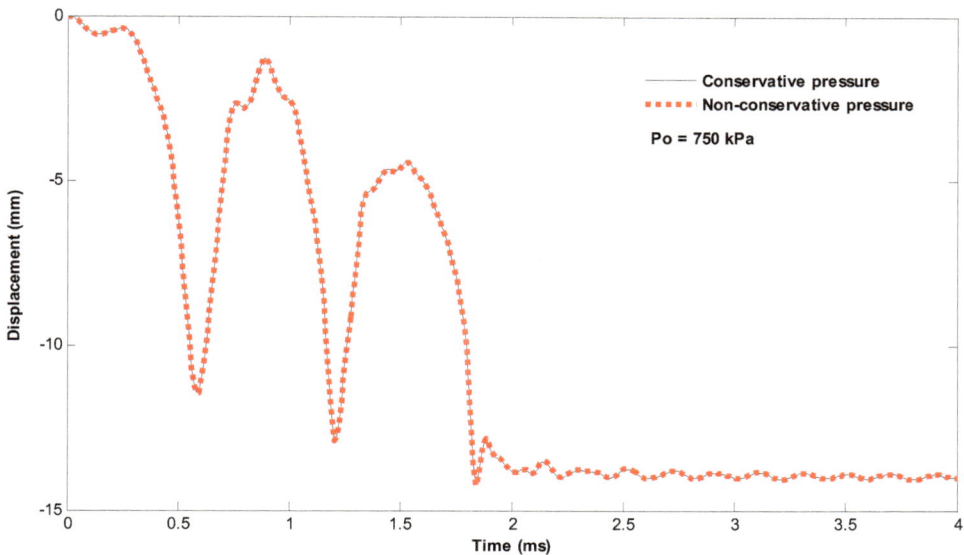

Figure 9.26(b): Nonlinear elastic responses of spherical cap at $P_0 = 750$ MPa.

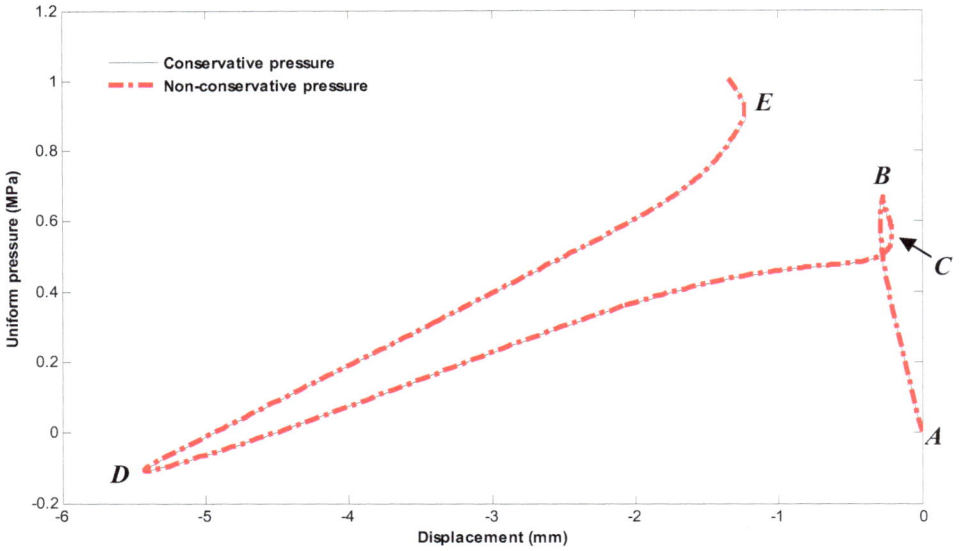

Figure 9.27(a): Relationship between load and nonlinear elastic displacement of hinged spherical cap.

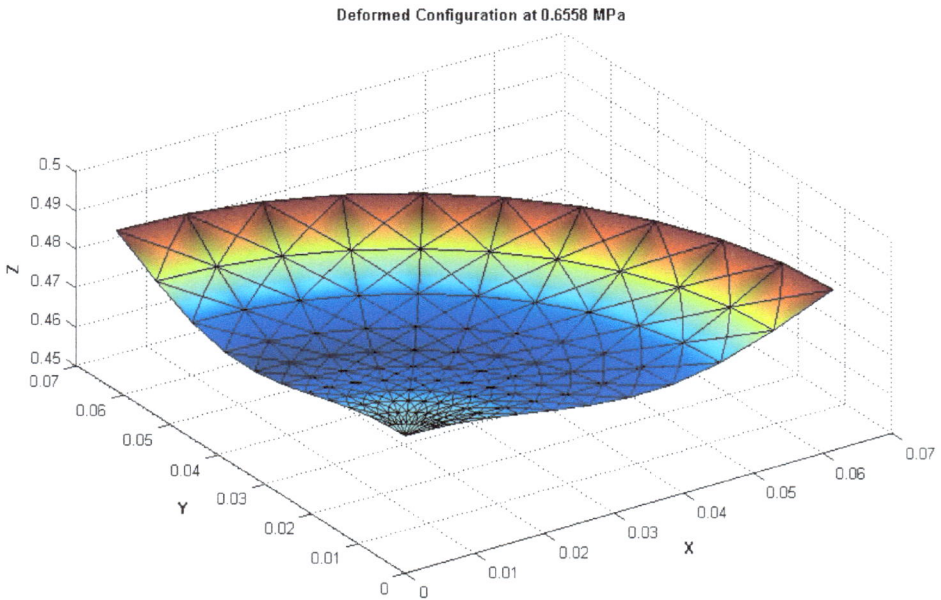

Figure 9.27(b): Deformed configuration of hinged spherical cap at 0.6558 MPa.

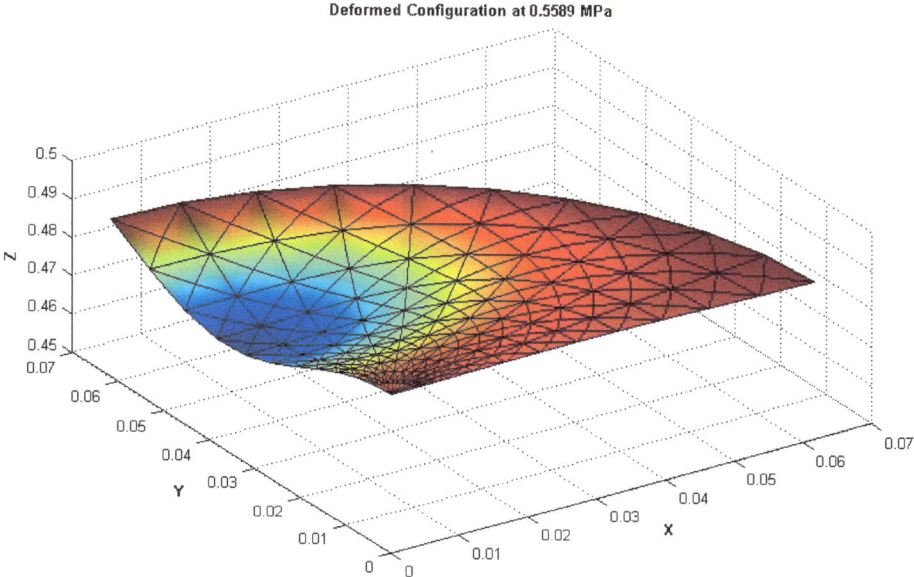

Figure 9.27(c): Deformed configuration of hinged spherical cap at 0.5589 MPa.

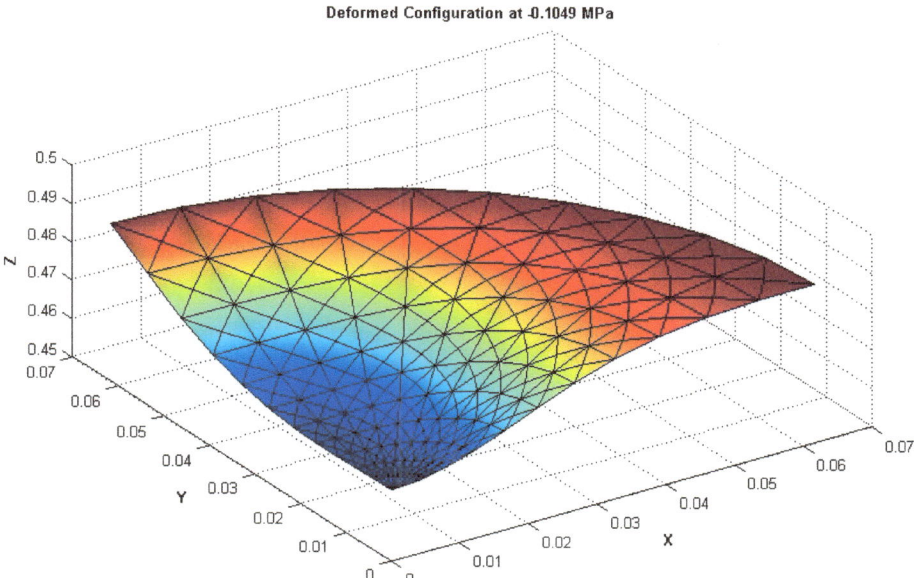

Figure 9.27(d): Deformed configuration of hinged spherical cap at -0.1049 MPa.

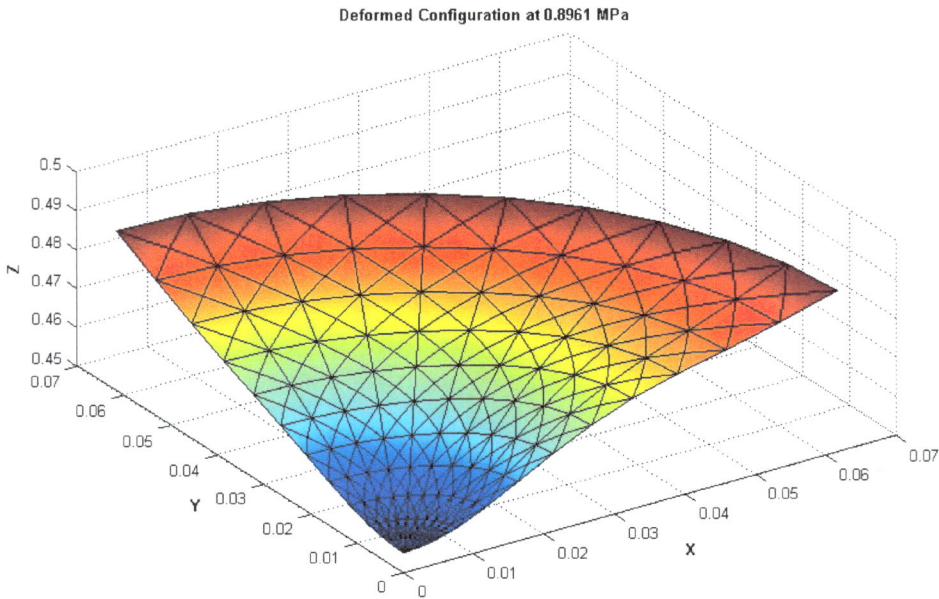

Deformed Configuration at 0.8961 MPa

Figure 9.27(e): Deformed configuration of hinged spherical cap at 0.8961 MPa.

9.7.2. Apex Load

The apex load $P(t)$ is of the impulse type which is identical to that considered in Sub-section 9.6.2. The load level P_0 is equal to 44.5 N (10 lb), 53.4 N (12 lb), 60.1 N (13 lb) and 66.75 N (15 lb). They are chosen after careful examination of the relationship between apex load and elastic-plastic displacement that is given in Fig. **9.28** which shows that at $P_0 = 62.14$ N (13.97 lb), snap-through takes place. Therefore, the chosen load levels range from lower to higher than the nonlinear buckling load. Computed apex responses are presented in Fig. **9.29**. It can be seen that, at the lower load levels (44.5 N and 53.4 N), the cap is displaced to the respective peak values. It then "settles onto" the respective permanent deformations. The load level of $P_0 = 60.1$ N causes dynamic snap-through. The cap seems to follow the path indicated in Fig. **9.28**, and snaps to approximately 2.5 mm and settles onto some permanent displacement. At the highest load level computed, the cap again experiences dynamic snap-through. However, the

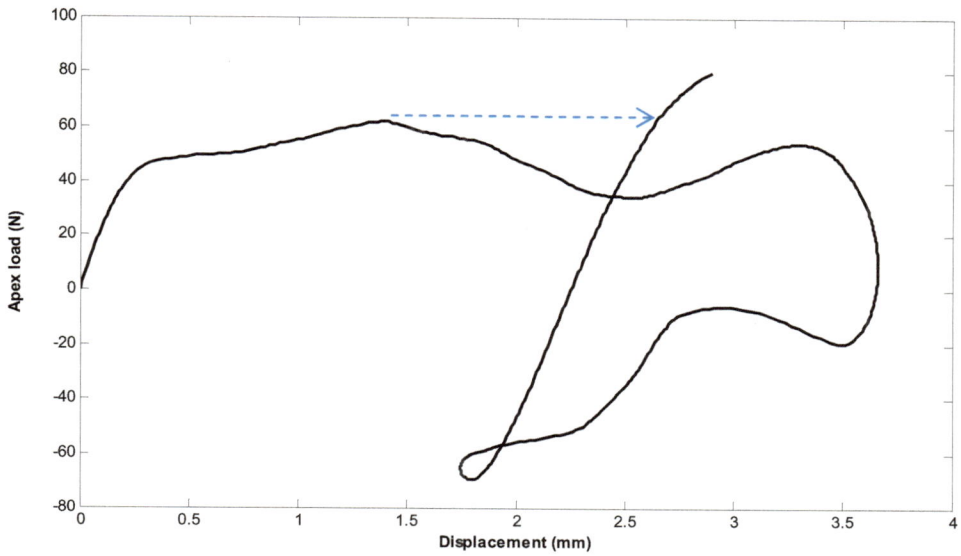

Figure 9.28: Relationship between apex load and elastic-plastic displacement of hinged spherical cap.

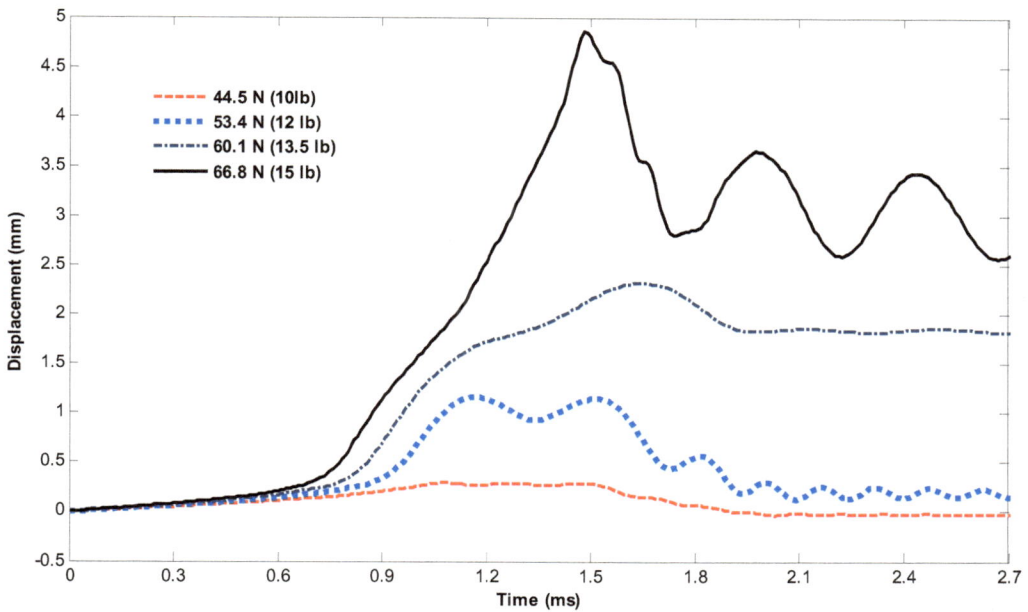

Figure 9.29: Nonlinear elasto-plastic responses of hinged spherical cap subjected to step load.

displacement to which the apex of cap snaps is not identifiable from Fig. **9.28**. Nor is the permanent mean displacement of approximately 3 mm. It should be kept in mind that Fig. **9.28** displays only the nonlinear static behaviour of the cap. Finally, it is noted that the time step size used is $\Delta t = 11.0$ μs. Displacements shown in Figs. **9.28** and **9.29** are in their absolute values. For all computation in this sub-section, the yield strength is $\sigma_Y = 137.9$ MPa (20 ksi).

REFERENCES

[9.1] J. Argyris, M. Papadrakakis, and Z.S. Mouroutis, "Nonlinear dynamics of shells with the triangular element TRIC", *Comp. Methods Appl. Mech. Engrg.*, vol. 192, pp. 3005-3038, July 2003.

[9.2] D. Kuhl, and E. Ramm, "Generalized energy–momentum method for nonlinear adaptive shell dynamics", *Comp. Methods Appl. Mech. Engrg.*, vol.178, pp. 343–366, August 1999.

[9.3] T. Liu, C. Zhao, Q. Li, and L. Zhang, "An efficient backward Euler time-integration method for nonlinear dynamic analysis of structures", *Comp. Struct.,* vol. 106-7, pp. 20-28, September 2012.

[9.4] U. Montag, W.B. Krätzig, and J. Soric, "Increasing solution stability for finite element modeling of elasto-plastic shell response", *Adv. Engineering Software*, vol. 30, pp. 607-619, September 1999.

[9.5] S.W. Key, and C.C. Hoff, "An improved constant membrane and bending stress shell element for explicit transient dynamics", *Comp. Methods Appl. Mech. Engrg.*, vol. 124, pp. 33-47, June 1995.

[9.6] C. Polat, and Y. Calayir, "Nonlinear static and dynamic analysis of shells of revolution", *Mechanics Res. Commun.*, vol 37, pp. 205-209, March 2010.

[9.7] C.W.S. To, and M.L. Liu, "Hybrid strain based three node flat triangular shell elements – II: numerical investigation of nonlinear problems", *Comp. Struct.*, vol. 54, pp. 1057-1076, March 1995.

[9.8] J. Oliver, and E. Onate, "A total Lagrangian formulation for the geometrically nonlinear analysis of structures using finite elements, – I: two-dimensional problems, shell and plate structures", *Int. J. Numer. Meth. Engng.*, vol. 20, pp. 2253-2281, 1984.

[9.9] K.J. Bathe, E.L. Wilson, and R.H. Iding, "NONSAP, a structural analysis program for static and dynamic response of nonlinear systems", Report No. SESM 74-3, Structural Engineering Laboratory, University of California, Berkeley, California, 1974.

[9.10] T. Hong, "Nonlinear, buckling and postbuckling analysis of shells of revolution", Ph.D. thesis, The Hong Kong Polytechnic University, Hong Kong, 1999.

[9.11] S. Kato, Y. Chiba, and I. Mutoh, "Secondary buckling analysis of spherical caps", *Struct. Engrg. Mechanics*, vol. 5, pp. 715-728, June 1997.

Author Index

INDEX

www.ingramcontent.com/pod-product-compliance
Lightning Source LLC
Chambersburg PA
CBHW050841220326
41598CB00006B/431